Agriculture Automation and Control

Series Editor

Qin Zhang, CPAAS, Washington State University, Prosser, WA, USA

The ultimate goal of agricultural research and technology development is to help farmers produce sufficient foods, feeds, fibers, or biofuels while at the same time, minimize the environmental impacts caused by these large scale activities. Automation offers a potential means by which improved productivity, resource optimization, and worker health and safety, can be accomplished. Although research on agricultural automation can be found in the published literature, there lacks a curated source of reference that is devoted to the unique characteristics of the agricultural system. This book series aims to fill the gap by bringing together scientists, engineers, and others working in these areas, and from around the world, to share their success stories and challenges. Individual book volume will have a focused theme and will be guest-edited by researchers/scientists renowned for their work within the respective sub-discipline.

Haoyu Niu • YangQuan Chen

Smart Big Data in Digital Agriculture Applications

Acquisition, Advanced Analytics, and Plant Physiology-informed Artificial Intelligence

Springer

Haoyu Niu
Texas A&M Institute of Data Science (TAMIDS)
Texas A&M University
College Station, TX, USA

YangQuan Chen
Department of Mechanical Engineering
University of California
Merced, CA, USA

ISSN 2731-3492 ISSN 2731-3506 (electronic)
Agriculture Automation and Control
ISBN 978-3-031-52644-2 ISBN 978-3-031-52645-9 (eBook)
https://doi.org/10.1007/978-3-031-52645-9

This Springer imprint is published by the registered company Springer Nature Switzerland AG
The registered company address is: Gewerbestrasse 11, 6330 Cham, Switzerland

Paper in this product is recyclable.

To our families.

Preface

In the dynamic realm of digital agriculture, the integration of big data acquisition platforms has sparked both curiosity and enthusiasm among researchers and agricultural practitioners. This book embarks on a journey to explore the intersection of artificial intelligence and agriculture, focusing on small unmanned aerial vehicles (UAVs), unmanned ground vehicles (UGVs), edge-AI sensors, and the profound impact they have on digital agriculture, particularly in the context of heterogeneous crops, such as walnuts, pomegranates, cotton, etc. For example, lightweight sensors mounted on UAVs, including multispectral and thermal infrared cameras, serve as invaluable tools for capturing high-resolution images. Their enhanced temporal and spatial resolutions, coupled with cost effectiveness and near-real-time data acquisition, position UAVs as an optimal platform for mapping and monitoring crop variability in vast expanses. This combination of data acquisition platforms and advanced analytics generates substantial datasets, necessitating a deep understanding of fractional-order thinking, which is imperative due to the inherent "complexity" and consequent variability within the agricultural process. Much optimism is vested in the field of artificial intelligence, such as machine learning (ML) and computer vision (CV), where the efficient utilization of big data to make it "smart" is of paramount importance in agricultural research. Central to this learning process lies the intricate relationship between plant physiology and optimization methods. The key to the learning process is the plant physiology and optimization method. Crafting an efficient optimization method raises three pivotal questions: (1) What represents the best approach to optimization? (2) How can we achieve a more optimal optimization? (3) Is it possible to demand "more optimal machine learning," exemplified by deep learning, while minimizing the need for extensive labeled data for digital agriculture?

In this book, the authors have explored the foundations of the plant physiology-informed machine learning (PPIML) and the principle of tail matching (POTM) framework. They elucidated their role in modeling, analyzing, designing, and managing complex systems based on big data in digital agriculture. Plant physiology embodies the intricacies of growth, and within this complex system, deterministic and stochastic dynamic processes coexist, influenced by external driving processes characterized and modeled using fractional calculus-based models. These insights better inform the development of complexity-informed machine learning (CIML) algorithms. To practically illustrate the application of these principles, data acquisition platforms, including low-cost UAVs, UGVs, and edge-AI sensors, were designed and built to demonstrate their reliability and robustness for remote and proximate sensing in agricultural applications. Research findings have shown that the PPIML, POTM, CIML, and the data acquisition platforms were reliable, robust, and smart tools for digital agricultural research across diverse scenarios, such as water stress detection, early detection of nematodes, yield estimation, and evapotranspiration (ET) estimation. The utilization of these tools holds the potential to significantly assist researchers and stakeholders in making informed decisions regarding crop management.

College Station, TX, USA
Merced, CA, USA
November 2023

Haoyu Niu
YangQuan Chen

Acknowledgments

The authors would like to thank Dr. Dong Wang for coordinating the USDA project and providing domain knowledge and comments. The authors also would like to thank Dr. Andreas Westphal for coordinating the UC Kearny project and providing the domain knowledge and comments on nematode detection. The authors wish to thank Dr. Bruce J. West for his wisdom sharing on "complexity matching" and "complexity synchronization" in recent decade. Thanks go to Prof. Mukesh Singhal, Prof. Wan Du, and Dr. Tiebiao Zhao. Their encouraging, critical, and constructive comments and suggestions increased the value of this monograph. Finally, many graduate and undergraduate researchers helped collect and process data: Dong Sen Yan, Stella Zambrzuski, Andreas Anderson, Allan Murillo, Christopher Currier, and Joshua Ahmed. We gratefully acknowledge the support of NVIDIA Corporation with the donation of the Titan X Pascal GPU used for this research. HN was supported by a Bayer Crop Science grant and a USDA NIFA grant entitled "Putting phenotypic and genotypic tools to work for improving walnut rootstocks." YC has been supported in part by the First UC CITRIS Aviation Prize (2022–2023) and the University of California Merced Climate Action Seed Fund 2023–2025 for CMERI, an NSF grant CBET-1856112 under the award entitled "INFEWS:T2: Saltwater Greenhouse System for Agricultural Drainage Treatment and Food Production," and an F3 R&D GSR Award (Farms Food Future Innovation Initiative (or F3), as funded by US Dept. of Commerce, Economic Development Administration Build Back Better Regional Challenge).

Contents

Acronyms

AI	Artificial Intelligence
ANN	Artificial Neural Network
ARS	Agricultural Sciences Center
BRDF	Bidirectional Reflectance Distribution Function
CIMIS	California Irrigation Management Information System
CIML	Complexity-Informed Machine Learning
CNNs	Convolutional Neural Networks
CRP	Calibrated Reflectance Panel
DEM	Digital Elevation Model
DLS	Downwelling Light Sensor
DN	Digital Number
DNNs	Deep Neural Networks
DOY	Day of Year
DTD	Dual Temperature Difference
ET	Evapotranspiration
FOV	Field of View
GPS	Global Positioning System
GPU	Graphics Processing Unit
HRMET	High Resolution Mapping of ET
ID	Identity
IoLT	Internet of Living Things
IR	Infrared
JPG	Joint Photographic Experts Group
LAI	Leaf Area Index
LDA	Linear Discriminant Analysis
MAE	Mean Absolute Error
METRIC	Mapping Evapotranspiration with Internalized Calibration
ML	Machine Learning
MLP	Multi-Layer Perceptron
NDVI	Normalized Difference Vegetation Index
NIST	National Institute of Standards and Technology

NIR	Near Infrared
OSEB	One-Source Energy Balance
PA	Precision Agriculture
PCA	Principal Component Analysis
PDF	Probability Distribution Function
POTM	Principle of Tail Matching
PPIML	Plant Physiology-Informed Machine Learning
QDA	Quadratic Discriminant Analysis
RGB	Red, Green, and Blue
RMSE	Root Mean Square Error
RSEB	Remote Sensing Energy Balance
SCN	Stochastic Configuration Network
SEBAL	Surface Energy Balance Algorithm for Land
SGD	Stochastic Gradient Descent
SVM	Support Vector Machine
SWIR	Short-Wave Infrared
TIR	Thermal Infrared
TSEB	Two-Source Energy Balance
TSEB-PT	Priestley-Taylor TSEB
UAVs	Unmanned Aerial Vehicles
UGVs	Unmanned Ground Vehicles
US	United States
USDA	United States Department of Agriculture
VIS	Visible

Part I
Why Big Data Is Not Smart Yet?

Chapter 1
Introduction

Abstract The introduction of this book sets the stage by exploring the motivation behind leveraging smart big data in the context of digital agriculture. It begins by elucidating the concept of smart big data and its relevance in transforming the agricultural landscape. The authors delve into the intersection of plant physiology and artificial intelligence, presenting it as a novel frontier in agricultural research. The integration of big data acquisition and advanced analytics is highlighted as a key catalyst for innovation in the field. The objectives and methods employed in the book are outlined, providing a roadmap for the readers to navigate the forthcoming content. Emphasis is placed on the unique contributions that the book brings to the intersection of digital agriculture, smart big data, and artificial intelligence. The chapter concludes with a detailed outline, offering a preview of the book's structure and the topics covered in each section. This comprehensive introduction serves as a foundation for the readers to grasp the significance of plant physiology-informed artificial intelligence in the realm of digital agriculture, providing a compelling rationale for the subsequent chapters. The References section ensures scholarly credibility and points readers toward additional resources for further exploration.

1.1 Motivation

The term "big data" started showing up in the early 1990s. The world's technological per capita capacity to store information has roughly doubled every 40 months since the 1980s [10]. Since 2012, there have been 2.5 exabytes (2.5×2^{60} bytes) of data generated every day [36]. According to data report predictions, there will be 163 zettabytes of data by 2025 [31]. Firican proposed ten characteristics (properties) of big data to prepare for both the challenges and advantages of big data initiatives in [7].

- **Volume:** the best known characteristic of big data; more than 90% of the whole data were created in the past couple of years.
- **Velocity:** the speed at which data are being generated
- **Variety:** processing structured, unstructured, and semi-structured data

H. Niu, Y. Chen, *Smart Big Data in Digital Agriculture Applications*, Agriculture Automation and Control, https://doi.org/10.1007/978-3-031-52645-9_1

- **Variability:** inconsistent speed of data loading, multitude of data dimensions, and the number of inconsistencies
- **Veracity:** confidence or trust in the data
- **Validity:** It refers to how accurate and correct the data are.
- **Vulnerability:** security concerns and data breaches
- **Volatility:** design policy for data currency, availability, and rapid retrieval of information when required
- **Visualization:** It develops new tools considering the complex relationships between the above properties.
- **Value:** the most important of the 10 Vs; a substantial value must be found.

With the development of big data technologies and high-performance computing, big data creates new opportunities for farmers and researchers to quantify, analyze, and better understand data intensive processes in digital agriculture. Big data can provide us information on the weather, irrigation management, pest management, fertilizer requirements, and so on. This enables farmers to make better decisions, such as what kind of crops to grow for better profitability and when to harvest.

However, big data technology also faces challenges that have been discussed and reviewed by many researchers [20, 23, 40]. For example, Zhang et al. pointed out three challenges faced by agricultural big data in [40], which were big data storage, big data analysis, and big data timeliness. Data storage could affect the efficiency of data analysis; they proposed using timeliness as a measurement standard based on the characteristics of agricultural big data. In [20], Gopal et al. proposed that how to obtain reliable data on farm management decision-making for both current conditions and under scenarios of changing biophysical and socioeconomic conditions is also one of the greatest challenges for big data applications in agriculture. Considering the complexity of agricultural datasets, multiple data models and algorithms are also needed in different procedures of big data processing. There are many challenges for traditional methods to extract meaningful information from big data [23], such as what is the optimal management zone for crops and what is the optimal zone size for soil sampling to analyze variability? The benefit of big data for digital agricultural applications remains elusive.

Therefore, the authors proposed the concept of smart big data for agricultural applications using machine learning (ML) algorithms, in which the variability analysis plays a key role. In this book, **variability** is the most important characteristic discussed for agricultural research. Variability refers to several properties of big data. First, it refers to the number of inconsistencies in the data, which need to be understood by using anomaly- and outlier-detection methods for any meaningful analytics to be performed. Second, variability can also refer to diversity [1, 25], resulting from disparate data types and sources, for example, healthy or unhealthy [14, 19]. Finally, variability can refer to multiple research topics [26].

1.1.1 What Is Smart Big Data in Digital Agriculture?

As mentioned earlier, big data technology, such as the Internet of Things (IoT) and wireless sensors, enabled researchers to solve complex agricultural problems [32]. By applying the sensors in the field, farmers can track valuable data for farm management, such as soil moisture, wind speed, air temperature, humidity, and so on. The amount of data can be huge and challenging to process timely. How to make big data "smarter" becomes necessary. Thus, the concept of smart big data analysis is proposed in this book. What do authors mean by "Smart"? Inspired by the Smart and Autonomous Systems (S&AS) proposed by National Science Foundation in 2018, the authors summarized "Smart" as (1) cognizant, (2) taskable, (3) reflective, (4) ethical, and (5) knowledge-rich. "Cognizant" exhibits high-level awareness beyond primitive actions, in support of persistent and long-term autonomy. "Taskable" can interpret high-level, possibly vague, instructions, translating them into concrete actions that are dependent on the particular context. "Reflective" can learn from their own experiences and those of other entities, such as other systems or humans, and from instruction or observation; they may exhibit self-aware and self-optimizing capabilities. "Ethical" should adhere to a system of societal and legal rules, taking those rules into account when making decisions. "Knowledge-rich" employs a variety of representation and reasoning mechanisms, such as semantic, probabilistic, and commonsense reasoning; are cognitively plausible; reason about uncertainty in decision making; and reason about the intentions of other entities in decision-making.

Smart big data in agricultural applications is an interdisciplinary research topic related to the extraction of meaningful information from plant physiology data, drawing techniques from a variety of fields, such as UAV image processing, deep learning, pattern recognition, high-performance computing, and statistics. Big data can then be filtered and becomes smart big data before being analyzed for insights, which leads to more efficient decision-making. Smart big data can be defined as big data that has been (1) collected, (2) preprocessed (such as cleaned, filtered, and prepared for data analysis), and (3) actionable information extraction in a smart way.

More and more researchers are gaining interest in smart big data in digital agricultural applications [9, 30, 41]. For example, Li and Niu proposed a design for smart agriculture using the big data and IoT in [17]. They optimized the data storage, data processing, and data mining procedures generated in the agricultural production process and used the k-means algorithm to study data mining. Based on the experimental results, the improved k-means-clustering method had an average reduction of 0.23 s in the total time and an average increase of 7.67% in the F metric value. In [35], Tseng et al. utilized the IoT devices to monitor the environmental factors on a farm. The experimental results demonstrated that farmers can gain a better understanding if a crop is appropriate for their farm by looking into factors such as temperature and soil moisture content. In [13], a big data analytic agricultural framework was developed to identify disease based on symptoms similarity, and a solution was suggested based on the high similarity. Although their framework is crop and location specific, it has a great potential to expand to more crops and areas in the future.

Researchers are trying all kinds of methods to turn the collected big data into smart big data to gain better understanding of our agricultural system. The authors believe that smart big data will be a core component of big data applications in digital agriculture, enabling stakeholders and researchers to identify patterns, make better decisions, and adapt to the new environment. Smart big data will also lay the foundation for agricultural data analysis.

1.1.2 Plant Physiology-Informed Artificial Intelligence: A New Frontier

Artificial intelligence (AI) represents a pivotal frontier in the realm of technology, changing how machines perceive and interact with the world. At the heart of AI, two integral branches stand out: Machine Learning and Computer Vision. Machine learning equips machines with the ability to learn from data, enabling them to recognize patterns, make decisions, and continuously improve their performance. Computer vision, on the other hand, enables machines to understand and interpret the visual world, akin to the human sense of sight. Through the amalgamation of Machine Learning and Computer Vision, AI systems can discern objects [46], navigate environments [16], and process visual information with astonishing accuracy. Together, these domains epitomize the potential of AI in transforming industries, from healthcare and autonomous vehicles to e-commerce and entertainment, and hold the promise of creating intelligent systems that can perceive and understand our world in ways previously thought impossible.

Machine Learning (ML) is the science (and art) of programming computers, so they can learn from data [8]. A more engineering-oriented definition was given by Tom Mitchell in 1997 [24], "A computer program is said to learn from experience E with respect to some task T and some performance measure P, if its performance on T, as measured by P, improves with experience E." In 2006, Hinton et al. trained a deep neural network (DNN) to recognize handwritten digits with an accuracy of more than 98% [11]. Since then, researchers are more and more interested in Deep Learning (DL), and this enthusiasm extends to many areas of ML, such as image processing [42, 44], natural language processing [12], and even digital agriculture [27, 28, 43].

Why do we need ML? In summary, ML algorithms can usually simplify a solution and perform better than traditional methods, which may require much more hand-tuning rules. Furthermore, there may not exist a right solution for the complex phenomena by traditional methods. The ML techniques can help explain that kind of complexity and can adapt to new data better. The ML algorithms can obtain the variability about the complex problems and big data [26]. There are many different types and ways for ML algorithms classification (Fig. 1.1). ML can be classified

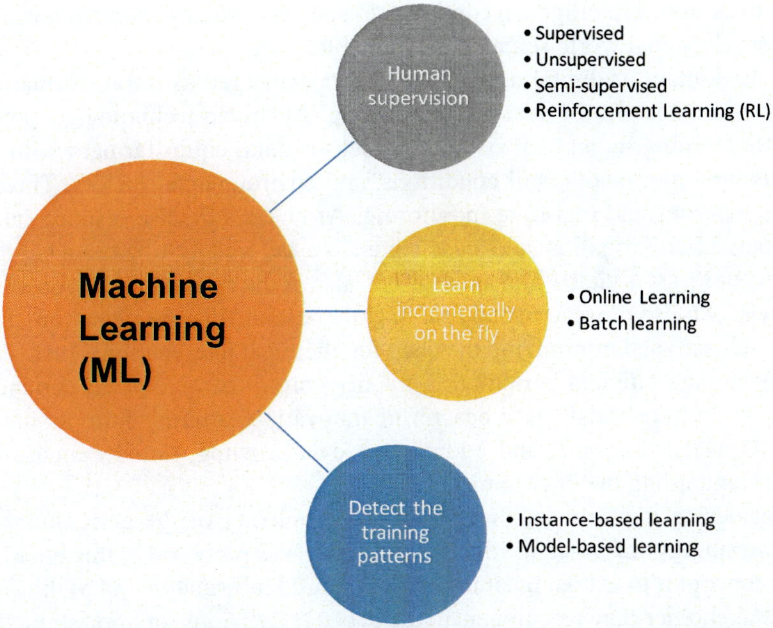

Fig. 1.1 The ML can be classified as supervised, unsupervised, semi-supervised, and Reinforcement Learning (RL) based on whether or not human supervision is included. According to whether or not the ML algorithms can learn incrementally on the fly, they can be classified into online and batch learning. Based on whether or not the ML algorithms detect the training data patterns and create a predictive model, the ML can be classified into instance-based and model-based learning

as supervised, unsupervised, semi-supervised, and Reinforcement Learning (RL) based on whether human supervision is included. According to whether or not the ML algorithms can learn incrementally on the fly, they can be classified into online and batch learning. Based on whether or not the ML algorithms detect the training data patterns and create a predictive model, the ML can be classified into instance-based and model-based learning [8].

Computer vision is the discipline of teaching machines to interpret and comprehend the visual world. It is the art of enabling computers to extract meaningful information from images and videos, much like the human ability to perceive and understand visual data. Over the years, computer vision has made remarkable strides, with applications spanning from facial recognition systems [3] and autonomous vehicles [6] to medical image analysis [5] and augmented reality [22]. Its significance is underscored by advancements in deep learning, particularly convolutional neural networks (CNNs), which have redefined the field. In 2012, a deep neural network known as AlexNet demonstrated an unprecedented ability to recognize objects in images, paving the way for a new era of image analysis [15]. Since then, computer vision has witnessed exponential growth, with diverse applications in fields such as healthcare, security, and industrial automation. As the demand for visual intelligence continues to soar, computer vision's relevance in our increasingly visual world remains indisputable.

In the domain of digital agriculture, AI has emerged as a transformative force, ushering in a new era of data-driven farming. AI-driven technologies, powered by machine learning, computer vision, and sensor data, equip farmers with valuable insights into crop health, soil conditions, and environmental factors. Through predictive analytics and real-time monitoring, AI enables precise resource allocation, from optimized irrigation and pest control to tailored fertilization [4]. This not only bolsters crop yields and quality but also reduces waste and conserves vital resources. AI also plays a pivotal role in early pest and disease detection, mitigating potential losses and minimizing the need for chemical interventions. In an era where sustainable and efficient agricultural practices are of paramount importance, AI in digital agriculture stands as a beacon of innovation, offering farmers the tools to make informed decisions and address the ever-growing global demand for food while safeguarding the environment.

Considering the volume, diversity, and complexity of the agricultural dataset, plant physiology-informed artificial intelligence was proposed in this book. The key of this concept is to extract meaningful agricultural information out of the big data to guide stakeholders and researchers to make better decisions for agriculture, in which the big data becomes "smart." Instead of training the ML or CV models directly, plant physiology knowledge will be added into the training process, which helps explain the complexity and model performance. When complexity is under scrutiny, it is fair that we ask what it means? At what point do investigators begin identifying a system, network, or phenomenon as complex [37, 38]? It seems that a clear and unified definition of complexity is still unknown for us to answer the following questions:

1. How can we characterize complexity?
2. What method should be used for the analysis of complexity in order to better understand real-world complex phenomena, such as the evapotranspiration of trees?

There is agreement among a significant fraction of the scientific community that when the distribution of the data associated with the process of interest is IPL, the phenomenon is complex. In the book by West and Grigolini [39], there is a table listing a sample of the empirical power laws and IPLs uncovered in the past two centuries. For example, in scale-free networks, the degree distributions follow an IPL in connectivity [2, 34], and in the processing of signals containing pink noise the power spectrum is IPL [29]. For other examples, such as the probability distribution function (PDF), the auto-correlation function (ACF) [18], allometry ($Y = aX^b$) [45], anomalous relaxation (evolving over time) [33], anomalous diffusion (mean squared dissipation versus time) [21], and self-similar, they can all be described by an IPL (see the details in Chap. 2).

1.1.3 Big Data Acquisition and Advanced Analytics

Smart big data involves the use of artificial intelligence to make big data acquisition and advanced analytics actionable and transform big data into insights and provides engagement capabilities for researchers and stakeholders. The smart big data acquisition and advanced analytics refer to the use of classification, conversion, extraction, and analysis methods to extract meaningful information from agricultural data. The acquisition and advanced analytics process generally contain the data preparation, the data analysis, and the result evaluation and explanation. Data preparation involves the agricultural data collection and integration using smart big data acquisition platforms, such as UAVs, Edge-AI sensors, and UGVs. Data analytics refers to examining the large dataset and extracting the useful information out of the raw dataset by using ML algorithms and tools, such as PyTorch, TensorFlow, OpenCV, etc. The result evaluation and explanation involves the verification of patterns or characteristics produced by data analytics.

1.2 The Book Objectives and Methods

Considering that smart big data is a new concept with a great potential in digital agricultural applications, the main objective of this book is developing a methodological framework for the plant physiology-informed artificial intelligence supported by (1) smart big data acquisition platforms, such as UAV, Edge-AI sensors, and UGV and (2) advanced data analytics, such as fractional-order thinking and artificial intelligence.

In order to accomplish the main objectives, the smart big data in digital agricultural applications will be grouped into the following specific parts:

1. Why Big Data Is Not Smart Now?
2. Smart Big Data Acquisition Platforms
3. Advanced Big Data Analytics, Plant Physiology-Informed Artificial Intelligence, and Fractional-Order Thinking
4. Toward Smart Big Data in Digital Agriculture

In the first part, the concept of smart big data is proposed and discussed in Chap. 1 to build the framework of smart big data applications for digital agriculture. The authors discuss the importance of smart big data and investigated the correlation between the smart big data, machine learning, and fractional dynamics in Chap. 2.

In the second part, smart big data acquisition platforms were mainly discussed. The authors propose a UAV platform for remote sensing data collection. A reliable image processing workflow is proposed. The challenges and opportunities for UAV image processing are also discussed in Chap. 3. In Chap. 4, the authors propose the concept of IoLT and presented several proximate sensors. The potential of UGV platforms for agriculture is briefly discussed in Chap. 5.

For the third part, the authors propose the concept of plant physiology-informed artificial intelligence and how to use advanced analytics and fractional-order thinking to make contributions. Many novice researchers in the field of digital agriculture often find it challenging to embark on using AI for agricultural applications, especially if they lack a background in computer science. Learning the intricacies of machine learning or computer vision can be time-consuming and overwhelming. In response to this common dilemma, we have dedicated Chap. 6 to a beginner-friendly tutorial that covers the fundamental aspects of a machine learning or computer vision workflow. This tutorial is designed to provide a solid starting point and ease the entry into the world of AI-driven agricultural solutions for those with limited prior knowledge in the field. In Chap. 7, a non-invasive proximate sensing method for early detection of nematodes is proposed. Microwave reflectance from walnut leaves was analyzed using ML algorithms to classify the nematode infection levels in the walnut roots. In Chap. 8, reliable tree-level ET estimation methods are proposed using the UAV high-resolution imagery, ML algorithms, and platforms, such as Python, MATLAB, Pytorch, and TensorFlow. In Chap. 9, individual tree-level water status inference is performed using the high-resolution UAV thermal imagery and complexity-informed machine learning. In Chap. 10, the authors propose a scale-aware pomegranate yield prediction method using UAV imagery and machine learning.

In the fourth part, the authors discuss an intelligent bugs mapping and wiping robot for farmers in Chap. 11, which has a great potential for pest management in the future. Then, the authors propose a non-invasive stem water potential monitoring method using proximate sensor and ML algorithms for a walnut orchard in Chap. 12 and a low-cost soil moisture monitoring method in Chap. 13. At the end, the authors draw conclusive remarks and discuss the future research plan in Chap. 14.

1.3 Book Contributions

The main contribution of the book is to lay the foundations of the Smart Big Data in Digital Agricultural Applications. A framework is created to enable the plant physiology-informed artificial intelligence using the smart big data acquisition platforms and advanced data analytics. Likewise, the following contributions can be obtained from this book:

- A developing framework for plant physiology-informed artificial intelligence adaptive to different kinds of trees and crops
- A set of data acquisition platforms, advanced analytics for smart big data applications in digital agriculture
- A beginner-friendly tutorial that covers the fundamental aspects of a machine learning or computer vision workflow to provide a solid starting point and ease the entry into the world of AI-driven agricultural solutions for those with limited prior knowledge in the field

- A Non-invasive method for early detection of nematodes in walnut using Edge-AI sensors
- Proposed reliable tree-level ET estimation methods using UAV and remote sensing sensors
- A concept of complexity-informed machine learning and its application for tree-level irrigation treatment inference
- Scale-aware yield estimation method using UAV thermal image and plant physiology-informed ML algorithms
- Proposed a UGV platform "iBMW" for pest management in agriculture
- Investigated potential Edge-AI sensors for future agricultural research

1.4 The Book Outline

The book is organized as follows: In Part I, the authors discuss the questions "Why big data is not smart now?" and "Why do we need smart big data for digital agriculture?" (Chap. 1). The objectives, methods, and contributions of this book are also listed in Chap. 1. In Chap. 2, and the authors present the correlation between the big data, machine learning, and the fractional dynamics. The authors try to answer "Why do big data and machine learning entail the fractional dynamics?"

In Part II, the smart big data acquisition platforms are presented and their applications for digital agriculture are introduced. Chapter 3 presents the UAV platforms and the remote sensing sensors mounted on them. The UAV image acquisition workflow is described in detail. Chapter 4 introduces the concept of IoLT and several Edge-AI sensors for agricultural applications. The potential of UGV platforms for agriculture is briefly discussed in Chap. 5.

In Part III, the main contribution of the authors' research work is discussed. Advanced big data analytics and fractional-order thinking are used for plant-informed machine learning. A beginner-friendly tutorial that covers the fundamental aspects of a machine learning or computer vision workflow is prepared in Chap. 6. It provides a solid starting point and eases the entry into the world of AI-driven agricultural solutions for those with limited prior knowledge in the field. In Chap. 7, a low-cost proximate sensing method for early detection of nematodes in walnut orchard is presented. Evapotranspiration estimation with small UAVs is mainly discussed in Chap. 8. Reliable tree-level ET estimation methods are also proposed in this chapter. In Chap. 9, individual tree-level water status inference is performed using the high-resolution UAV thermal imagery and complexity-informed machine learning. In Chap. 10, the authors propose a scale-aware pomegranate yield prediction method using UAV imagery and machine learning.

In Part IV, the authors discuss an intelligent bugs mapping and wiping robot for farmers in Chap. 11, which has a great potential for pest management in the future. Then, the authors propose a non-invasive stem water potential monitoring method using proximate sensor and ML algorithms for a walnut orchard in Chap. 12 and a low-cost soil moisture monitoring method in Chap. 13. At the end, the authors draw conclusive remarks and discuss the future research in Chap. 14.

References

1. Arabas, J., Opara, K.: Population diversity of non-elitist evolutionary algorithms in the exploration phase. IEEE Trans. Evol. Comput. **24**(6), 1050–1062 (2019)
2. Barabási, A.L., Albert, R.: Emergence of scaling in random networks. Science **286**(5439), 509–512 (1999)
3. Barnouti, N.H., Al-Dabbagh, S.S.M., Matti, W.E.: Face recognition: a literature review. Int. J. Appl. Inform. Syst. **11**(4), 21–31 (2016)
4. Chandra, R., Collis, S.: Digital agriculture for small-scale producers: challenges and opportunities. Commun. ACM **64**(12), 75–84 (2021)
5. Elyan, E., Vuttipittayamongkol, P., Johnston, P., Martin, K., McPherson, K., Jayne, C., Sarker, M.K., et al.: Computer vision and machine learning for medical image analysis: recent advances, challenges, and way forward. Artif. Intell. Surg. **2** (2022)
6. Faisal, A., Kamruzzaman, M., Yigitcanlar, T., Currie, G.: Understanding autonomous vehicles. J. Transp. Land Use **12**(1), 45–72 (2019)
7. Firican, G.: The 10 Vs of Big Data (2017). https://tdwi.org/articles/2017/02/08/10-vs-of-big-data.aspx
8. Géron, A.: Hands-on Machine Learning with Scikit-Learn, Keras, and TensorFlow: Concepts, Tools, and Techniques to Build Intelligent Systems. O'Reilly Media (2019)
9. Granda-Cantuna, J., Molina-Colcha, C., Hidalgo-Lupera, S.E., Valarezo-Varela, C.D.: Design and implementation of a wireless sensor network for precision agriculture operating in API mode. In: 2018 International Conference on eDemocracy & eGovernment (ICEDEG), pp. 144–149. IEEE, Piscataway (2018)
10. Hilbert, M., López, P.: The world's technological capacity to store, communicate, and compute information. Science **332**(6025), 60–65 (2011)
11. Hinton, G.E., Osindero, S., Teh, Y.W.: A fast learning algorithm for deep belief nets. Neural Comput. **18**(7), 1527–1554 (2006)
12. Hirschberg, J., Manning, C.D.: Advances in natural language processing. Science **349**(6245), 261–266 (2015)
13. Kaur, R., Garg, R., Aggarwal, H.: Big data analytics framework to identify crop disease and recommendation a solution. In: 2016 International Conference on Inventive Computation Technologies (ICICT), vol. 2, pp. 1–5. IEEE, Piscataway (2016)
14. Ko, M., Stark, B., Barbadillo, M., Chen, Y.: An evaluation of three approaches using Hurst estimation to differentiate between normal and abnormal HRV. In: Proceedings of the International Design Engineering Technical Conferences and Computers and Information in Engineering Conference. American Society of Mechanical Engineers (2015)
15. Krizhevsky, A., Sutskever, I., Hinton, G.E.: Imagenet classification with deep convolutional neural networks. In: Advances in Neural Information Processing Systems, vol. 25 (2012)
16. Levine, S., Shah, D.: Learning robotic navigation from experience: principles, methods and recent results. Philos. Trans. Roy. Soc. B **378**(1869), 20210,447 (2023)
17. Li, C., Niu, B.: Design of smart agriculture based on big data and Internet of things. Int. J. Distrib. Sensor Netw. **16**(5), 1550147720917,065 (2020)
18. Li, M.: Modeling autocorrelation functions of long-range dependent teletraffic series based on optimal approximation in Hilbert space—a further study. Appl. Math. Model. **31**(3), 625–631 (2007)
19. Li, N., Cruz, J., Chien, C.S., Sojoudi, S., Recht, B., Stone, D., Csete, M., Bahmiller, D., Doyle, J.C.: Robust efficiency and actuator saturation explain healthy heart rate control and variability. Proc. Natl. Acad. Sci. **111**(33), E3476–E3485 (2014)
20. Maya-Gopal, P., Chintala, B.R., et al.: Big data challenges and opportunities in agriculture. Int. J. Agric. Environ. Inform. Syst. **11**(1), 48–66 (2020)
21. Metzler, R., Klafter, J.: The random walk's guide to anomalous diffusion: a fractional dynamics approach. Phys. Rep. **339**(1), 1–77 (2000)

22. Minaee, S., Liang, X., Yan, S.: Modern augmented reality: applications, trends, and future directions (2022). arXiv preprint arXiv:2202.09450
23. Mintert, J.R., Widmar, D., Langemeier, M., Boehlje, M., Erickson, B.: The challenges of precision agriculture: is big data the answer? Tech. rep., Southern Agricultural Economics Association (SAEA) Annual Meeting, San Antonio, Texas, February 6–9, 2015 (2016)
24. Mitchell, T.M.: Machine Learning. McGraw-Hill, New York (1997)
25. Nakahira, Y., Liu, Q., Sejnowski, T.J., Doyle, J.C.: Diversity-enabled sweet spots in layered architectures and speed-accuracy trade-offs in sensorimotor control (2019). arXiv preprint arXiv:1909.08601
26. Niu, H., Chen, Y., West, B.J.: Why do big data and machine learning entail the fractional dynamics? Entropy **23**(3), 297 (2021)
27. Niu, H., Wang, D., Chen, Y.: Estimating actual crop evapotranspiration using deep stochastic configuration networks model and UAV-based crop coefficients in a pomegranate orchard. In: Autonomous Air and Ground Sensing Systems for Agricultural Optimization and Phenotyping V. International Society for Optics and Photonics (2020)
28. Niu, H., Zhao, T., Wang, D., Chen, Y.: A UAV resolution and waveband aware path planning for onion irrigation treatments inference. In: 2019 International Conference on Unmanned Aircraft Systems (ICUAS), pp. 808–812. IEEE, Piscataway (2019)
29. Pesquet-Popescu, B., Pesquet, J.C.: Synthesis of bidimensional α-stable models with long-range dependence. Signal Process. **82**(12), 1927–1940 (2002)
30. Rajeswari, S., Suthendran, K., Rajakumar, K.: A smart agricultural model by integrating IoT, mobile and cloud-based big data analytics. In: 2017 International Conference on Intelligent Computing and Control (I2C2), pp. 1–5. IEEE, Piscataway (2017)
31. Reinsel, D., Gantz, J., Rydning, J.: Data age 2025: The evolution of data to life-critical don't focus on big data; focus on the data that's big. International Data Corporation (IDC) White Paper (2017)
32. Sourav, A., Emanuel, A.: Recent trends of big data in precision agriculture: a review. In: IOP Conference Series: Materials Science and Engineering, vol. 1096.1, p. 012081. IOP Publishing (2021)
33. Sun, H., Chen, Y., Chen, W.: Random-order fractional differential equation models. Signal Process. **91**(3), 525–530 (2011)
34. Sun, W., Li, Y., Li, C., Chen, Y.: Convergence speed of a fractional order consensus algorithm over undirected scale-free networks. Asian J. Control **13**(6), 936–946 (2011)
35. Tseng, F.H., Cho, H.H., Wu, H.T.: Applying big data for intelligent agriculture-based crop selection analysis. IEEE Access **7**, 116965–116974 (2019)
36. Ward, J.S., Barker, A.: Undefined by data: a survey of big data definitions (2013). arXiv preprint arXiv:1309.5821
37. West, B.J.: Sir Isaac Newton stranger in a strange land. Entropy **22**(11), 1204 (2020)
38. West, B.J., Geneston, E.L., Grigolini, P.: Maximizing information exchange between complex networks. Phys. Rep. **468**(1–3), 1–99 (2008)
39. West, B.J., Grigolini, P.: Complex Webs: Anticipating the Improbable. Cambridge University Press, Cambridge (2010)
40. Zhang, H., Wei, X., Zou, T., Li, Z., Yang, G.: Agriculture big data: research status, challenges and countermeasures. In: International Conference on Computer and Computing Technologies in Agriculture, pp. 137–143. Springer, Berlin (2014)
41. Zhang, P., Zhang, Q., Liu, F., Li, J., Cao, N., Song, C.: The construction of the integration of water and fertilizer smart water saving irrigation system based on big data. In: 2017 IEEE International Conference on Computational Science and Engineering (CSE) and IEEE International Conference on Embedded and Ubiquitous Computing (EUC), vol. 2, pp. 392–397. IEEE, Piscataway (2017)
42. Zhao, T., Koumis, A., Niu, H., Wang, D., Chen, Y.: Onion irrigation treatment inference using a low-cost hyperspectral scanner. In: Multispectral, Hyperspectral, and Ultraspectral Remote Sensing Technology, Techniques and Applications VII. International Society for Optics and Photonics (2018)

43. Zhao, T., Niu, H., de la Rosa, E., Doll, D., Wang, D., Chen, Y.: Tree canopy differentiation using instance-aware semantic segmentation. In: 2018 ASABE Annual International Meeting. American Society of Agricultural and Biological Engineers (2018)
44. Zhao, T., Yang, Y., Niu, H., Wang, D., Chen, Y.: Comparing U-Net convolutional network with mask R-CNN in the performances of pomegranate tree canopy segmentation. In: Multispectral, Hyperspectral, and Ultraspectral Remote Sensing Technology, Techniques and Applications VII, vol. 10780, p. 107801J. International Society for Optics and Photonics (2018)
45. Zhao, Z., Guo, Q., Li, C.: A fractional model for the allometric scaling laws. Open Appl. Math. J. **2**(1), 26–30 (2008)
46. Zou, Z., Chen, K., Shi, Z., Guo, Y., Ye, J.: Object detection in 20 years: a survey. Proceedings of the IEEE (2023)

Chapter 2
Why Do Big Data and Machine Learning Entail the Fractional Dynamics?

Abstract Chapter 2 explores the fundamental question of why big data and machine learning inherently involve fractional dynamics. The investigation unfolds through an exploration of fractional calculus (FC) and fractional-order thinking (FOT), shedding light on their relevance in understanding the intricate dynamics of complex systems. The chapter delves into the concept of complexity and inverse power laws (IPLs), establishing a connection between heavy-tailed distributions and fractional dynamics. Various heavy-tailed distributions are examined in the context of their implications for machine learning in diverse applications. The discussion extends to the interplay between big data, variability, and fractional calculus, incorporating topics such as the Hurst parameter, fractional Gaussian noise (fGn), etc. The chapter progresses to elucidate the concept of optimal machine learning and optimal randomness, distinguishing between derivative-free methods and gradient-based methods. Additionally, it explores the contributions of the control community to machine learning. A detailed case study on optimal randomness for Stochastic Configuration Network (SCN) with heavy-tailed distributions is presented, encompassing an introduction, SCN with heavy-tailed probability density functions (PDFs), a regression model, parameter tuning, and a practical application involving MNIST handwritten digit classification.

2.1 Fractional Calculus (FC) and Fractional-Order Thinking (FOT)

Fractional calculus (FC) is the quantitative analysis of functions using non-integer-order integration or differentiation, where the order can be a real number, a complex number, or even a function of a variable. The first recorded query regarding the meaning of a non-integer order differentiation appeared in a letter written in 1695 by Guillaume de l'Hôpital to Gottfried Wilhelm Leibniz, who at the same time as Isaac Newton, but independently of him, co-invented the infinitesimal calculus [141]. Numerous contributors have provided definitions for fractional derivatives and integrals [140] since then, and the theory along with the applications of FC has been expanded greatly over the centuries [1, 111, 119]. In more recent decades, the

H. Niu, Y. Chen, *Smart Big Data in Digital Agriculture Applications*, Agriculture Automation and Control, https://doi.org/10.1007/978-3-031-52645-9_2

concept of fractional dynamics has merged and gained followers in the statistical and chemical physics communities [66, 113, 136]. For example, optimal image processing has improved through the use of fractional-order differentiation and fractional-order partial differential equations as summarized in Chen et al. [17, 18, 162]. Anomalous diffusion was described using fractional diffusion equations in [92, 123].

Recently, big data and machine learning (ML) are two of the hottest topics in applied scientific research, and they are closely related to each other. To better understand them, we also need fractional dynamics, as well as fractional-order thinking (FOT). Section 2.4 is devoted to the discussion of the relationships between big data, variability, and fractional dynamics, as well as to fractional-order data analytics (FODA) [126]. The topics touched on in this section include the Hurst parameter [42, 91], fractional Gaussian noise (fGn), fractional Brownian motion (fBm), the fractional autoregressive integrated moving average (FARIMA) [78], the formalism of continuous-time random walk (CTRW) [94], unmanned aerial vehicles (UAVs), and digital agriculture [77]. In Sect. 2.5, how to learn efficiently (optimally) for ML algorithms is investigated. The key to developing an efficient learning process is the method of optimization. Thus, it is important to design an efficient or perhaps optimal optimization method. The derivative-free methods, and the gradient-based methods, such as the Nesterov accelerated gradient descent (NAGD) [98], are discussed.

FOT is a way of thinking using FC. For example, there are non-integers between the integers; between logic 0 and logic 1, there is fuzzy logic [165]; compared to integer-order splines, there are fractional-order splines [139]; between the high-order integer moments, there are non-integer-order moments, etc. FOT has been entailed by many research areas, for example, self-similar [25, 120], scale-free or scale-invariant, power law, long-range-dependence (LRD) [15, 109], and $1/f^\alpha$ noise [53, 159]. The terms porous media, particulate, granular, lossy, anomaly, disorder, soil, tissue, electrodes, biology [7], nano, network, transport, diffusion, and soft matters are also intimately related to FOT. However, in the present section, the author mainly discusses **complexity and inverse power laws (IPLs)**.

2.2 Complexity and Inverse Power Laws (IPLs)

When studying complexity, it is fair to ask, what does it mean to be complex? When do investigators begin identifying a system, network, or phenomenon as being complex [155, 156]? There is an agreement among a significant fraction of the scientific community that when the distribution of the data associated with the process of interest obeys an IPL, the phenomenon is complex; see Fig. 2.1. On the left side of the figure, the complexity "bow tie" [26, 30, 31, 169] is the phenomenon of interest, thought to be a complex system. On the right side of the figure is the spectrum of system properties associated with IPL probability density functions (PDFs): the system has one or more of the properties of being scale-

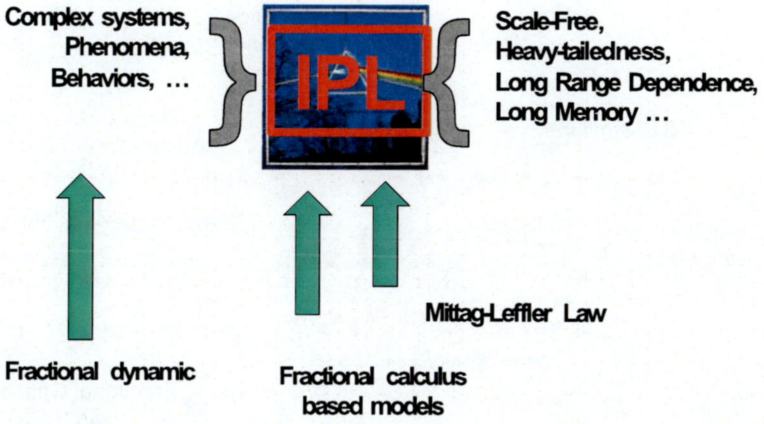

Fig. 2.1 Inverse power law (complexity "bow tie"): On the left are the systems of interest that are complex. In the center panel, an aspect of the empirical data is characterized by an inverse power law (IPL). The right panel lists the potential properties associated with systems with data that have been processed and yield an IPL property. See the text for more details

free, having a heavy tail, having a long-range dependence, and/or having a long memory [46, 127]. In the book by West and Grigolini [157], there is a table listing a sample of the empirical power laws and IPLs uncovered in the past two centuries. For example, in scale-free networks, the degree distributions follow an IPL in connectivity [8, 132]; in the processing of signals containing pink noise, the power spectrum follows an IPL [109]. For other examples, such as the probability density function (PDF), the autocorrelation function (ACF) [72], allometry ($Y = aX^b$) [172], anomalous relaxation (evolving over time) [130], anomalous diffusion (mean squared dissipation versus time) [92], and self-similarity can all be described by the IPL "bow tie" depicted in Fig. 2.1.

The power law is usually described as:

$$f(x) = ax^k, \tag{2.1}$$

when k is negative, $f(x)$ is an IPL. One important characteristic of this power law is scale invariance [63] determined by:

$$f(cx) = a(cx)^k = c^k f(x) \propto f(x). \tag{2.2}$$

Note that when x is the time, the scaling depicts a property of the system dynamics. However, when the system is stochastic, the scaling is a property of the PDF (or correlation structure) and is a constraint on the collective properties of the system.

FC is entailed by complexity, since an observable phenomenon represented by a fractal function has integer-order derivatives that diverge. Consequently, for the complexity characterization and regulation, we ought to use the fractional dynamics point of view because the fractional derivative of a fractal function is finite. Thus,

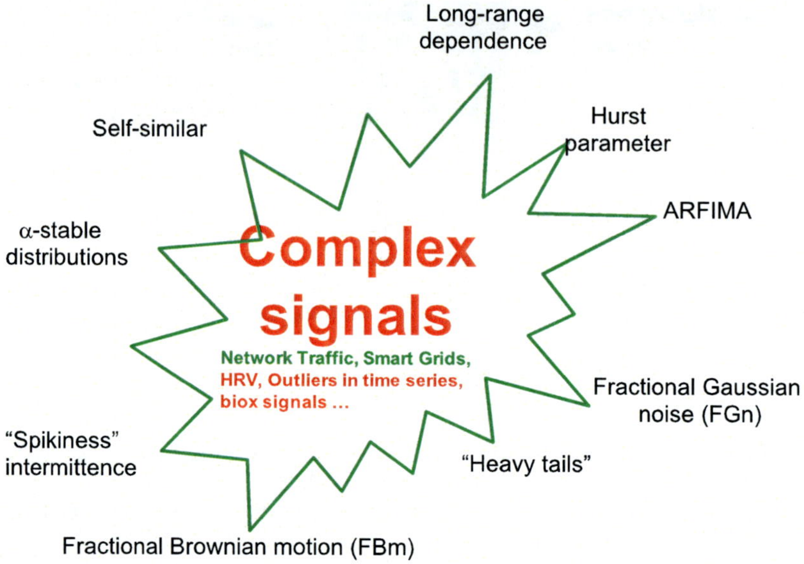

Fig. 2.2 Complex signals (IPL): Here, the signal generated by a complex system is depicted. Exemplars of the systems are given as the potential properties arising from the systems' complexity

complex phenomena, no matter whether they are natural or carefully engineered, ought to be described by fractional dynamics. Phenomena in complex systems in many cases should be analyzed using FC-based models, where mathematically, the IPL is actually the "Mittag–Leffler law" (MLL), which asymptotically becomes an IPL (Fig. 2.2), known as heavy-tail behavior.

When an IPL results from processing data, one should think about how the phenomena can be connected to the FC. In [44], Gorenflo and Mainardi explained the role of the FC in generating stable PDFs by generalizing the diffusion equation to the one with fractional order. For the Cauchy problem, they considered the space-fractional diffusion equation:

$$\frac{\partial u}{\partial t} = D(\alpha)\frac{\partial^{\alpha} u}{\partial |x|^{\alpha}}, \tag{2.3}$$

where $-\infty < x < \infty$, $t \geq 0$ with $u(x, 0) = \delta(x)$, $0 < \alpha \leq 2$, and $D(\alpha)$ is a suitable diffusion coefficient. The fractional derivative in the diffusion variable is of the Reisz–Feller form, defined by its Fourier transform to be $|k|^a$. For the signaling problem, they considered the so-called time-fractional diffusion equation [88]:

$$\frac{\partial^{2\beta} u}{\partial t^{2\beta}} = D(\beta)\frac{\partial^{2} u}{\partial x^{2}}, \tag{2.4}$$

where $x \geq 0, t \geq 0$ with $u(0, t) = \delta(t), 0 < \beta < 1$, and $D(\beta)$ is a suitable diffusion coefficient. Equation (2.4) has also been investigated in [83–85]. Here, the Caputo fractional derivative in time is used.

There are rich forms in stochasticity [95], for example, heavy-tailedness, which corresponds to fractional-order master equations [76]. In Sect. 2.3, heavy-tailed distributions are discussed.

2.3 Heavy-Tailed Distributions

In probability theory, heavy-tailed distributions are PDFs whose tails do not decay exponentially [5]. Consequently, they have more weight in their tails than does an exponential distribution. In many applications, it is the right tail of the distribution that is of interest, but a distribution may have a heavy left tail, or both tails may be heavy. Heavy-tailed distributions are widely used for modeling in different disciplines, such as finance [10], insurance [2], and medicine [115]. The distribution of a real-valued random variable X is said to have a heavy right tail if the tail probabilities $P(X > x)$ decay more slowly than those of any exponential distribution:

$$\lim_{x \to \infty} \left(\frac{P(X > x)}{e^{-\lambda x}} \right) = \infty, \tag{2.5}$$

for every $\lambda > 0$ [117]. For the heavy left tail, an analogous definition can be constructed [38]. Typically, there are three important subclasses of heavy-tailed distributions: fat-tailed, long-tailed, and subexponential distributions.

2.3.1 Lévy Distribution

A Lévy distribution, named after the French mathematician Paul Lévy, can be generated by a random walk whose steps have a probability of having a length determined by a heavy-tailed distribution [99]. As a fractional-order stochastic process with heavy-tailed distributions, a Lévy distribution has better computational characteristics [51]. A Lévy distribution is stable and has a PDF that can be expressed analytically, although not always in closed-form. The PDF of Lévy flight [163] is:

$$p(x, \mu, \gamma) = \begin{cases} \frac{\sqrt{\frac{\gamma}{2\pi}}}{e^{-\frac{\gamma}{2(x-\mu)}} (x-\mu)^{3/2}}, & x > \mu, \\ 0, & x \leq \mu, \end{cases} \tag{2.6}$$

where μ is the location parameter and γ is the scale parameter. In practice, the Lévy distribution is updated by

$$Lévy(\beta) = \frac{u}{|\nu|^{1/\beta}}, \tag{2.7}$$

where u and ν are random numbers generated from a normal distribution with a mean of 0 and standard deviation of 1 [164]. The stability index β ranges from 0 to 2. Furthermore, it is interesting to note that the well-known Gaussian and Cauchy distributions are special cases of the Lévy PDF when the stability index is set to 2 and 1, respectively.

2.3.2 Mittag–Leffler Distribution

The Mittag–Leffler PDF [54] for the time interval between events can be written as a mixture of exponentials with a known PDF for the exponential rates:

$$E_\theta(-t^\theta) = \int_0^\infty exp(-\mu t)g(\mu)d\mu, \tag{2.8}$$

with a weight for the rates given by:

$$g(\mu) = \frac{1}{\pi}\frac{\sin(\theta\pi)}{\mu^{1+\theta} + 2\cos(\theta\pi)\mu + \mu^{1-\theta}}. \tag{2.9}$$

The most convenient expression for the random time interval was proposed by Jayakumar [60]:

$$\tau_\theta = -\gamma_t\left(\ln u \frac{\sin(\theta\pi)}{\tan(\theta\pi v)} - \cos(\theta\pi)\right)^{1/\theta}, \tag{2.10}$$

where u, $v \in (0,1)$ are independent uniform random numbers, γ_t is the scale parameter, and τ_θ is the Mittag-Leffler random number. In [150], Wei et al. used the Mittag-Leffer distribution for improving the Cuckoo Search algorithm, which showed an improved performance.

2.3.3 Weibull Distribution

A random variable is described by a Weibull distribution if it has a PDF function F:

$$F(x) = e^{-(x/k)^{\lambda_w}}, \tag{2.11}$$

where $k > 0$ is the scale parameter and $\lambda_w > 0$ is the shape parameter [116]. If the shape parameter is $\lambda_w < 1$, the Weibull distribution is determined to be heavy tailed.

2.3.4 Cauchy Distribution

A random variable is described by a Cauchy PDF if its cumulative distribution is [35, 61]:

$$F(x) = \frac{1}{\pi}\arctan\left(\frac{2(x-\mu_c)}{\sigma}\right) + \frac{1}{2}, \tag{2.12}$$

where μ_c is the location parameter and σ is the scale parameter. Cauchy distributions are examples of fat-tailed distributions, which have been empirically encountered in a variety of areas including physics, earth sciences, economics, and political science [80]. Fat-tailed distributions include those whose tails decay like an IPL, which is a common point of reference in their use in the scientific literature [6]:

2.3.5 Pareto Distribution

A random variable is said to be described by a Pareto PDF if its cumulative distribution function is

$$F(x) = \begin{cases} 1 - (\frac{b}{x})^a, x \geq b, \\ 0, x < b, \end{cases} \tag{2.13}$$

where $b > 0$ is the scale parameter and $a > 0$ is the shape parameter (Pareto's index of inequality) [40].

2.3.6 The α-Stable Distribution

A PDF is said to be stable if a linear combination of two independent random variables, each with the same distribution, has the same distribution for the conjoined variable. This PDF is also called the Lévy α-stable distribution [71, 89]. Since the normal distribution, Cauchy distribution, and Lévy distribution all have the above property, one can consider them to be special cases of stable distributions. Stable distributions have $0 < \alpha \leq 2$, with the upper bound corresponding to the normal distribution, and $\alpha = 1$, to the Cauchy distribution. The PDFs have undefined variances for $\alpha < 2$, and undefined means for $\alpha \leq 1$. Although their PDFs do not admit a closed-form formula in general, except in special cases, they decay with an IPL tail called the IPL index, which determines the behavior of the PDF. As the IPL index gets smaller, the PDF acquires a heavier tail.

2.3.7 *Mixture Distributions*

A mixture distribution is derived from a collection of other random variables. First, a random variable is selected by chance from the collection according to given probabilities of selection. Then, the value of the selected random variable is realized. The mixture PDFs are complicated in terms of simpler PDFs, which provide a good model for certain datasets. The different subsets of the data can exhibit different characteristics. Therefore, the mixed PDFs can effectively characterize the complex PDFs of certain real-world datasets. In [81], a robust stochastic configuration network (SCN) based on a mixture of Gaussian and Laplace PDFs was proposed. Thus, Gaussian and Laplace distributions are mentioned in this section for comparison purposes.

2.3.7.1 Gaussian Distribution

A random variable X has a Gaussian distribution with the mean μ_G and variance σ_G^2 ($-\infty < \mu_G < \infty$ and $\sigma_G > 0$) if X has a continuous distribution for which the PDF is as follows [129]:

$$f(x|\mu_G, \sigma_G^2) = \frac{1}{(2\pi)^{1/2}\sigma_G} e^{-\frac{1}{2}(\frac{x-\mu_G}{\sigma_G})^2}, \; for \; -\infty < x < \infty. \tag{2.14}$$

2.3.7.2 Laplace Distribution

The PDF of the Laplace distribution can be written as follows [81]:

$$F(x|\mu_l, \eta) = \frac{1}{(2\eta^2)^{1/2}} e^{(-\frac{\sqrt{2}|x-\mu_l|}{\eta})}, \tag{2.15}$$

where μ_l and η represent the location and scale parameters, respectively.

2.4 Big Data, Variability, and FC

The term "big data" started showing up in the early 1990s. The world's technological per capita capacity to store information has roughly doubled every 40

Table 2.1 The 10 Vs of big data

Characteristics	Description
Volume	Best known characteristic of big data, more than 90 percent of the whole data were created in the past couple of years.
Velocity	The speed at which data are being generated.
Variety	Processing structured, unstructured, and semistructured data.
Variability	Inconsistent speed of data loading, multitude of data dimensions, and number of inconsistencies.
Veracity	Confidence or trust in the data.
Validity	Refers to how accurate and correct the data are.
Vulnerability	Security concerns, data breaches.
Volatility	Design policy for data currency, availability, and rapid retrieval of information when required.
Visualization	Develop new tools considering the complex relationships between the above properties.
Value	The most important of the 10 Vs, substantial value must be found.

months since the 1980s [56]. Since 2012, there have been 2.5 exabytes (2.5×2^{60} bytes) of data generated every day [148]. According to data report predictions, there will be 163 zettabytes of data by 2025 [114]. Firican proposed, in [37], ten characteristics (properties) of big data to prepare for both the challenges and advantages of big data initiatives (Table 2.1). The authors believe **variability** is the most important characteristic being discussed. Variability refers to several properties of big data. First, it refers to the number of inconsistencies in the data, which need to be understood by using anomaly- and outlier-detection methods for any meaningful analytics to be performed. Second, variability can also refer to diversity [4, 97], resulting from disparate data types and sources, for example, healthy or unhealthy [67, 74]. Finally, variability can refer to multiple research topics (Table 2.2).

Considering variability, Xunzi (312 BC–230 BC), who was a Confucian philosopher, made a useful observation: "Throughout a thousand acts and ten thousand changes, his way remains one and the same [59]." Therefore, we ask: what is the "one and the same" for big data? This is the **variability**, which refers to the behavior of the dynamic system. The ancient Greek philosopher Heraclitus (535 BC–475 BC) also realized the importance of variability, prompting him to say: "The only thing that is constant is change"; "It is in changing that we find purpose"; "Nothing

Table 2.2 Variability in multiple research topics

Topics	Description
1. Climate variability	Changes in the components of the climate system and their interactions.
2. Genetic variability	Measurements of the tendencies of individual genotypes between regions.
3. Heart rate variability	Physiological phenomenon where the time interval between heart beats varies.
4. Human variability	Measurements of the characteristics, physical or mental, of human beings.
5. Spatial variability	Measurements at different spatial points exhibit different values.
6. Statistical variability	A measure of dispersion in statistics.

endures but change"; "No man ever steps in the same river twice, for it is not the same river and he is not the same man."

Heraclitus actually recognized the (fractional-order) dynamics of the river without modern scientific knowledge (in nature). After 2000 years, the integer-order calculus was invented by Sir Issac Newton and Gottfried Wilhelm Leibniz, whose main purpose was to quantify that change [9, 12]. From then, scientists started using integer-order calculus to depict dynamic systems, differential equations, modeling, etc. In the 1950s, Scott Blair, who first introduced the FC into rheology, pointed out that the integer-order dynamic view of change is only for our own "convenience" (a little bit selfish). In other words, denying fractional calculus is equivalent to denying the existence of non-integers between the integers!

Blair said [134]: "We may express our concepts in Newtonian terms if we find this convenient but, if we do so, we must realize that we have made a translation into a language which is foreign to the system which we are studying (1950)."

Therefore, variability exists in big data as big data is generated from complex systems. However, how do we realize the modeling, analysis, and design (MAD) for the variability in big data within complex systems? We need fractional calculus! In other words, big data are at the nexus of complexity and FC. Thus, we first proposed fractional-order data analytics (FODA) in 2015. Metrics based on using the fractional-order signal processing techniques should be used for quantifying the generating dynamics of observed or perceived variability [126] inherent in the generating complex systems.

2.4.1 Hurst Parameter, fGn, and fBm

The Hurst parameter or Hurst exponent (H) was proposed for the analysis of the long-term memory of time series. It was originally developed to quantify the long-term storage capacity of reservoirs for the Nile river's volatile rain and drought conditions more than a half century ago [42, 91]. To date, the Hurst parameter has also been used to measure the intensity of long-range dependence (LRD) in time series [21], which requires accurate modeling and forecasting. The self-similarity and the estimation of the statistical parameters of LRD have commonly been investigated recently [128]. The Hurst parameter has also been used for characterizing the LRD process [21, 131]. A LRD time series is defined as a stationary process that has long-range correlations if its covariance function $C(n)$ decays slowly as:

$$\lim_{n\to\infty} \frac{C(n)}{n^{-\alpha}} = c, \tag{2.16}$$

where $0 < \alpha < 1$, which relates to the Hurst parameter according to $\alpha = 2 - 2H$ [110, 121]. The parameter c is a finite, positive constant. When the value of n is large, $C(n)$ behaves as the IPL c/n^{α} [48]. Another definition for an LRD process is that the weakly stationary time series $X(t)$ is said to be LRD if its power spectral density (PSD) follows:

$$f(\lambda) \sim C_f |\lambda|^{-\beta}, \tag{2.17}$$

as $\lambda \to 0$, for a given $C_f > 0$ and a given real parameter $\beta \in (0,1)$, which corresponds to $H = (1+\beta)/2$ [22]. When $0 < H < 0.5$, it indicates that the time intervals constitute a negatively correlated process. When $0.5 < H < 1$, it indicates that time intervals constitute a positively correlated process. When $H = 0.5$, it indicates that the process is uncorrelated.

Two of the most common LRD processes are fBm [27] and fGn [68]. The fBm process with $H (0 < H < 1)$ is defined as:

$$B_H(t) = \frac{1}{\Gamma(H+1/2)} \left\{ \int_{-\infty}^{0} \left[(t-s)^{H-1/2} - (-s)^{H-1/2} \right] dW(s) \right. \\ \left. + \int_0^t (t-s)^{H-1/2} dW(s) \right\}, \tag{2.18}$$

where W denotes a Wiener process defined on $(-\infty, \infty)$ [90]. The fGn process is the increment sequences of the fBm process, defined as:

$$X_k = Y(k+1) - Y(k), \tag{2.19}$$

where $Y(k)$ is a fBm process [106].

2.4.2 Fractional Lower-Order Moments (FLOMs)

The FLOM is based on α-stable PDFs. The PDFs of an α-stable distribution decay in the tails more slowly than a Gaussian PDF does. Therefore, for sharp spikes or occasional bursts in signals, an α-stable PDF can be used for characterizing signals more accurately than Gauss-distributed signals [20]. Thus, the FLOM plays an important role in impulsive processes [79], equivalent to the role played by the mean and variance in a Gaussian processes. When $0 < \alpha \leq 1$, the α-stable processes have no finite first- or higher order moments; when $1 < \alpha < 2$, the α-stable processes have a first-order moment and all the FLOMs with moments of fractional order that is less than 1. The correlation between the FC and FLOM was investigated in [23, 24]. For the Fourier-transform pair $p(x)$ and $\phi(\mu)$, the latter is the characteristic function and is the Fourier transform of the PDF; a complex FLOM can have complex fractional lower orders [23, 24]. A FLOM-based fractional power spectrum includes a covariation spectrum and a fractional low-order covariance spectrum [87]. FLOM-based fractional power spectrum techniques have been successfully used in time-delay estimation [87].

2.4.3 Fractional Autoregressive Integrated Moving Average (FARIMA) and Gegenbauer Autoregressive Moving Average (GARMA)

A continuous-time linear time-invariant (LTI) system can be characterized using a linear difference equation, which is known as an autoregression and moving average (ARMA) model [118, 124]. The process X_t of ARMA(p, q) is defined as:

$$\Phi(B)X_t = \Theta(B)\epsilon_t, \tag{2.20}$$

where ϵ_t is white Gaussian noise (wGn), and B is the backshift operator, $BX_t = X_{t-1}$. However, the ARMA model can only describe a short-range dependence (SRD) property. Therefore, based on the Hurst parameter analysis, more suitable models, such as FARIMA [52, 125] and fractional integral generalized autoregressive conditional heteroscedasticity (FIGARCH) [75], were designed to more accurately analyze the LRD processes. The most important feature of these models is the long-memory characteristic. The FARIMA and FIGARCH can capture both the short- and the long-memory nature of time series. For example, the FARIMA process X_t is usually defined as [13]:

$$\Phi(B)(1 - B)^d X_t = \Theta(B)\epsilon_t, \tag{2.21}$$

where $d \in (-0.5, 0.5)$, and $(1 - B)^d$ is a fractional-order difference operator. The locally stationary long-memory FARIMA model has the same equation as that of Eq. (2.21), except that d becomes d_t, which is a time-varying parameter [11]. The locally stationary long-memory FARIMA model captures the local self-similarity of the system.

The generalized locally stationary long-memory process FARIMA model was investigated in [11]. For example, a generalized FARIMA model, which is called the Gegenbauer autoregressive moving average (GARMA), was introduced in [47]. The GARMA model is defined as:

$$\Phi(B)(1 - 2uB + B^2)^d X_t = \Theta(B)\epsilon_t, \tag{2.22}$$

where $u \in [-1, 1]$, which is a parameter that can control the frequency at which the long memory occurs. The parameter d controls the rate of decay of the autocovariance function. The GARMA model can also be extended to the so-called k-factor GARMA model, which allows for long-memory behaviors to be associated with each of k frequencies (Gegenbauer frequencies) or seasonalities in the interval [0, 0.5] [160].

2.4.4 *Continuous-Time Random Walk (CTRW)*

The CTRW model was proposed by Montroll and Weiss as a generalization of diffusion processes to describe the phenomenon of anomalous diffusion [94]. The basic idea is to calculate the PDF for the diffusion process by replacing the discrete steps with continuous time, along with a PDF for step lengths and a waiting-time PDF for the time intervals between steps. Montroll and Weiss applied random intervals between the successive steps in the walking process to account for local structure in the environment, such as traps [154]. The CTRW has been used for modeling multiple complex phenomena, such as chaotic dynamic networks [167]. The correlation between CTRW and diffusion equations with fractional time derivatives has also been established [57]. Meanwhile, time-space fractional diffusion equations can be treated as CTRWs with continuously distributed jumps or continuum approximations of CTRWs on lattices [45].

2.4.5 *Unmanned Aerial Vehicles (UAVs) and Digital Agriculture*

As a new remote-sensing platform, researchers are more and more interested in the potential of small UAVs for digital agriculture [16, 28, 29, 43, 100–104, 133, 166], especially for heterogeneous crops, such as vineyards and orchards [170, 171]. Mounted on UAVs, lightweight sensors, such as RGB cameras, multispectral cameras, and thermal infrared cameras, can be used to collect high-resolution

Fig. 2.3 Normalized difference vegetation index (NDVI) mapping of pomegranate trees

images. The higher temporal and spatial resolutions of the images, relatively low operational costs, and nearly real-time image acquisition make the UAVs an ideal platform for mapping and monitoring the variability of crops and trees. UAVs can create big data and demand the FODA due to the "complexity" and, thus, variability inherent in the life process. For example, Fig. 2.3 shows the normalized difference vegetation index (NDVI) mapping of a pomegranate orchard at a USDA ARS experimental field. Under different irrigation levels, the individual trees can show strong variability during the analysis of water stress. Life is complex! Thus, it entails variability, which as discussed above, in turn, entails fractional dynamics. UAVs can then become "Tractor 2.0" for farmers in digital agriculture.

2.5 Optimal Machine Learning and Optimal Randomness

Machine learning (ML) is the science (and art) of programming computers so they can learn from data [41]. A more engineering-oriented definition was given by Tom Mitchell in 1997 [93], "A computer program is said to learn from experience E with respect to some task T and some performance measure P, if its performance on T, as measured by P, improves with experience E."

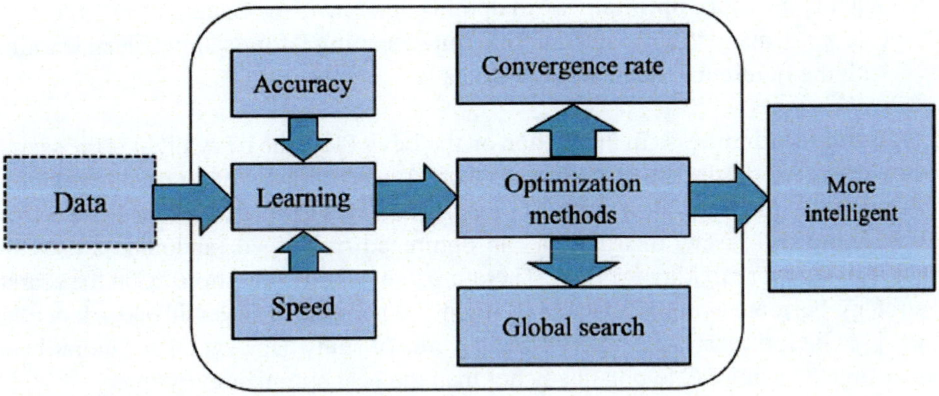

Fig. 2.4 Data analysis in nature

Most ML algorithms perform training by solving optimization problems that rely on first-order derivatives (Jacobians), which decide whether to increase or decrease weights. For huge speed boosts, faster optimizers are being used instead of the regular gradient descent optimizer. For example, the most popular boosters are momentum optimization [112], Nesterov accelerated gradient (NAGD) [98], AdaGrad [32], RMSProp [137], and Adam optimization [65]. The second-order (Hessian) optimization methods usually find the solutions with faster rates of convergence but with higher computational costs. Therefore, the answer to the following question is important: (1) what is a more optimal ML algorithm? (2) What if the derivative is fractional order instead of integer order? In this section, we discuss some applications of fractional-order gradients to optimization methods in machine learning algorithms and investigate the accuracy and convergence rates.

As mentioned in the big data section, there is a huge amount of data in human society and nature. During the learning process of ML, we care not only about the speed but also the accuracy of the data the machine is learning (Fig. 2.4). The learning algorithm is important; otherwise, the data labeling and other labor costs will exhaust people beyond their abilities. When applying the artificial intelligence (AI) to an algorithm, a strong emphasis is now on "artificial," only followed weakly by "intelligence." Therefore, the key to ML is what optimization methods are being applied, such that "artificial" part is less emphasized, e.g., labeling work can be minimized. The convergence rate and global searching are two important parts of the optimization method.

Reflection ML is, today, a hot research topic and will probably remain so into the near future. How a machine can learn efficiently (optimally) is always important. The key for the learning process is the optimization method. Thus, in designing an efficient optimization method, it is necessary to answer the following three questions:

- What is the optimal way to optimize?

- What is the **more optimal** way to optimize that involves fractional calculus?
- Can we demand **"more optimal machine learning,"** for example, deep learning with the minimum/smallest labeled data?

Optimal Randomness In the section on the Lévy PDF, the Lévy flight is the search strategy for food the albatross has developed over millions of years of evolution. Admittedly, this is a slow optimization procedure [144]. From this perspective, we should call "Lévy distribution" an optimized or learned randomness used by albatrosses for searching for food. Therefore, we pose the question: "Can the search strategy be more optimal than Lévy flight?" The answer is yes if one adopts the FC [168]! Optimization is a very complex area of study. However, few studies have investigated using FC to obtain a better than the best optimization strategy.

2.5.1 *Derivative-Free Methods*

For derivative-free methods, there are single agent search and swarm-based search methods (Fig. 2.5). Exploration is often achieved by randomness or random numbers in terms of some predefined PDFs. Exploitation uses local information such as gradients to search local regions more intensively, and such intensification can enhance the rate of convergence. Thus, a question was posed: what is the optimal randomness? Wei et al. [151] investigated the optimal randomness in a swarm-based search. Four heavy-tailed PDFs have been used for sample path analysis (Fig. 2.6). Based on the experimental results, the randomness-enhanced cuckoo search (CS) algorithms [150, 152, 153] can identify the unknown specific parameters of a fractional-order system with better effectiveness and robustness. The randomness-enhanced CS algorithms can be considered as a promising tool for solving real-world complex optimization problems. The reason is that optimal randomness is applied with fractional-order noise during the exploration, which is

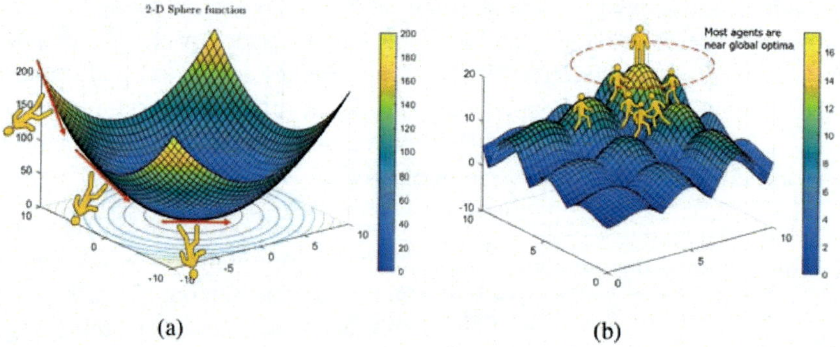

Fig. 2.5 The 2-D Alpine function for derivative-free methods. (**a**) Single agent search. (**b**) Swarm-based search methods

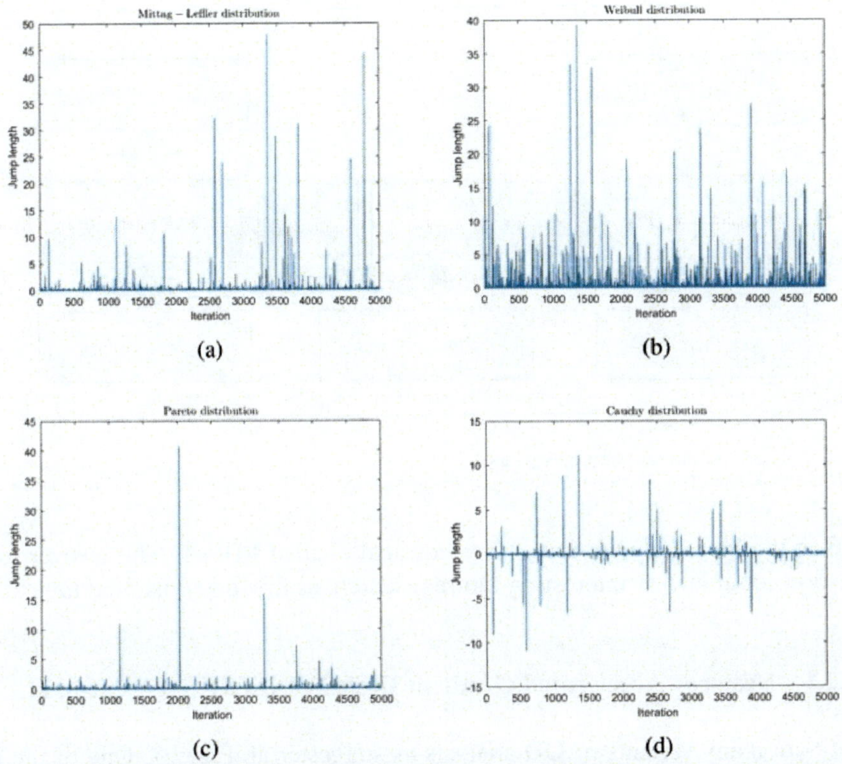

Fig. 2.6 Sample paths. Wei et al. investigated the optimal randomness in a swarm-based search. Four heavy-tailed PDFs were used for sample path analysis. The long steps, referring to the jump length, frequently happened for all distributions, which showed strong heavy-tailed performance. (**a**) Mittag-Leffler distribution. (**b**) Weibull distribution. (**c**) Pareto distribution. (**d**) Cauchy distribution

more optimal than the "optimized PSO," CS. The fractional-order noise refers to those with the stable PDFs [44]. In other words, when we are discussing optimal randomness, we are discussing fractional calculus!

2.5.2 *The Gradient-Based Methods*

The gradient descent (GD) is a very common optimization algorithm, which can find the optimal solutions by iteratively tweaking parameters to minimize the cost function. The stochastic gradient descent (SGD) randomly selects times during the training process. Therefore, the cost function bounces up and down, decreasing on average, which is good for escaping from local optima. Sometimes, noise is purposely added into the GD method, and usually, such noise follows a Gaussian

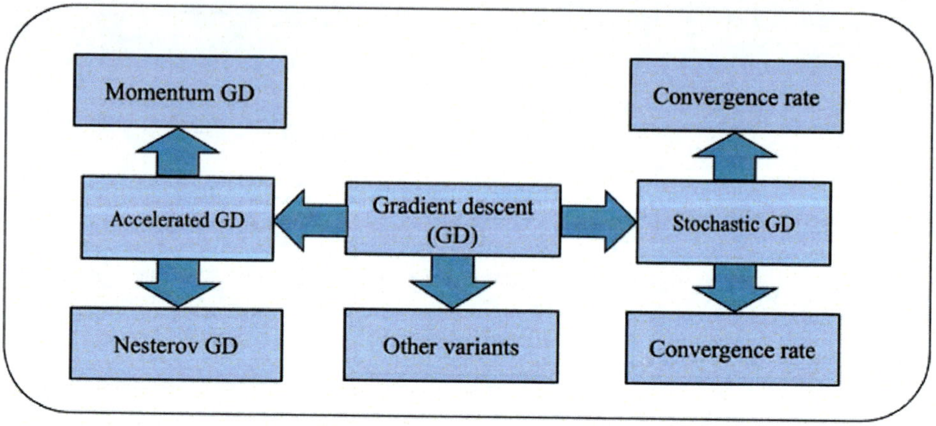

Fig. 2.7 Gradient descent and its variants

PDF in the literature. We ask, "why not heavy-tailed PDFs"? The answer to this question could lead to interesting future research, as first demonstrated in [151].

2.5.2.1 Nesterov Accelerated Gradient Descent (NAGD)

There are many variants of GD analysis as suggested in Fig. 2.7. One of the most popular methods is the NAGD [98]:

$$\begin{cases} y_{k+1} = a y_k - \mu \nabla f(x_k), \\ x_{k+1} = x_k + y_{k+1} + b y_k, \end{cases} \tag{2.23}$$

where by setting $b = -a/(1+a)$, one can derive the NAGD. When $b = 0$, one can derive the momentum GD. The NAGD can also be formulated as:

$$\begin{cases} x_k = y_{k-1} - \mu \nabla f(y_{k-1}), \\ y_k = x_k + \frac{k-1}{k+2}(x_k - x_{k-1}). \end{cases} \tag{2.24}$$

Set $t = k\sqrt{\mu}$, and one can, in the continuous limit, derive the corresponding differential equation:

$$\ddot{X} + \frac{3}{t}\dot{X} + \nabla f(X) = 0. \tag{2.25}$$

The main idea of Jordan's work [158] is to analyze the iteration algorithm in the continuous-time domain. For differential equations, one can use the Lyapunov or variational method to analyze the properties; for example, the convergence rate is $O(\frac{1}{t^2})$. One can also use the variational method to derive the master differential equation for an optimization method, such as the least action principle [36], Hamilton's variational principle [50] and the quantum-mechanical path integral approach [55]. Wilson et al. [158] built a Euler–Lagrange function to derive the following equation:

$$\ddot{X}_t + 2\gamma \dot{X}_t + \frac{\gamma^2}{\mu} \nabla f(X_t) = 0, \tag{2.26}$$

which is in the same form as the master differential equation of NAGD.

Jordan's work revealed that one can transform an iterative (optimization) algorithm to its continuous-time limit case, which can simplify the analysis (Laypunov methods). One can directly design a differential equation of motion (EOM) and then discretize it to derive an iterative algorithm (variational method). The key is to find a suitable Lyapunov functional to analyze the stability and convergent rate. The new exciting fact established by Jordan is that optimization algorithms can be systematically synthesized using Lagrangian mechanics (Euler–Lagrange) through EOMs.

Thus, is there an optimal way to optimize using optimization algorithms stemming from Eq. (2.26)? Obviously, why not an equation such as Eq. (2.26) of fractional order? Considering the $\dot{X}_t$ as $X_t^{(\alpha)}$, it will provide us with more research possibilities, such as the fractional-order calculus of variation (FOCV) and fractional-order Euler–Lagrange (FOEL) equation. For the SGD, optimal randomness using the fractional-order noises can also offer better than the best performance, similarly shown by Wei et al. [151].

2.6 What Can the Control Community Offer to ML?

In the IFAC 2020 World Congress Pre-conference Workshop, Eric Kerrigan proposed "The Three Musketeers" that the control community can contribute to ML community [64]. These are the IMP [39], the Nu-Gap metric [142], and model discrimination [143]. Herein, we focused on the IMP. Kashima et al. [62] transferred the convergence problem of numerical algorithms into a stability problem of a discrete-time system. An et al. [3] explained that the commonly used SGD-momentum algorithm in ML is a PI controller and designed a PID algorithm. Motivated by An et al. [3] but differently from M. Jordan's work, we proposed designing and analyzing the algorithms in the S or Z domain. Remember that GD is a first-order algorithm:

$$x_{k+1} = x_k - \mu \nabla f(x_k), \tag{2.27}$$

where $\mu > 0$ is the step size (or learning rate). Using the Z transform, one can achieve:

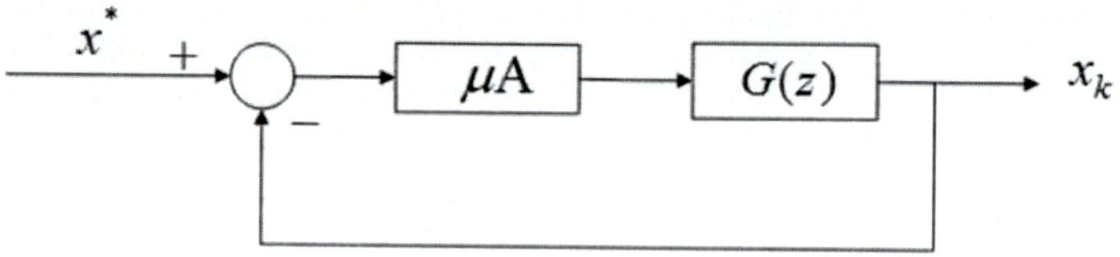

Fig. 2.8 The integrator model (embedded in $G(z)$). The integrator in the forward loop eliminates the tracking steady-state error for a constant reference signal (internal model principle (IMP))

$$X(z) = \frac{\mu}{z-1}[-\nabla f(x_k)]_z. \tag{2.28}$$

Approximate the gradient around the extreme point x^*, and one can obtain:

$$\nabla f(x_k) \approx A(x_k - x^*), \text{ with } A = \nabla^2 f(x^*). \tag{2.29}$$

For the plain GD in Fig. 2.8, we have $G(z) = 1/(z-1)$, which is an integrator. For fractional-order GD (FOGD), the updating term of x_k in Eq. (2.27) can be treated as a filtered gradient signal. In [33], Fan et al. shared similar thoughts: "Accelerating the convergence of the moment method for the Boltzmann equation using filters." The integrator in the forward loop eliminates the steady-state tracking error for a constant reference signal according to the internal model principle (IMP). Similarly, the GD momentum (GDM) designed to accelerate the conventional GD, which is popularly used in ML, can be analyzed using Fig. 2.8 by:

$$\begin{cases} y_{k+1} = \alpha y_k - \mu \nabla f(x_k), \\ x_{k+1} = x_k + y_{k+1}, \end{cases} \tag{2.30}$$

where y_k is the accumulation of the history gradient and $\alpha \in (0, 1)$ is the rate of the moving average decay. Using the Z transform for the update rule, one can derive:

$$\begin{cases} zY(z) = \alpha Y(z) - \mu[\nabla f(x_k)]_z, \\ zX(z) = X(z) + zY(z). \end{cases} \tag{2.31}$$

Then, after some algebra, one obtains the following equation:

$$X(z) = \frac{\mu z}{(z-1)(z-\alpha)}[-\nabla f(x_k)]_z. \tag{2.32}$$

For the GD momentum, we have $G(z) = \frac{z}{(z-1)(z-\alpha)}$ in Fig. 2.8, with an integrator in the forward loop. The GD momentum is a second-order ($G(z)$) algorithm with an additional pole at $z = \alpha$ and one zero at $z = 0$. The "second-order" refers to the order of $G(z)$, which makes it different from the algorithm using the $Hessian$ matrix information. Moreover, NAGD can be simplified as:

$$\begin{cases} y_{k+1} = x_k - \mu \nabla f(x_k), \\ x_{k+1} = (1-\lambda) y_{k+1} + \lambda y_k, \end{cases} \tag{2.33}$$

where μ is the step size and λ is a weighting coefficient. Using the Z transform for the update rule, one can derive:

$$\begin{cases} zY(z) = X(z) - \mu[\nabla f(x_k)]_z, \\ zX(z) = (1-\lambda)zY(z) + \lambda Y(z). \end{cases} \tag{2.34}$$

Different from the GD momentum, and after some algebra, one obtains:

$$X(z) = \frac{-(1-\lambda)z - \lambda}{(z-1)(z+\lambda)} \mu[\nabla f(x_k)]_z = \frac{z + \frac{\lambda}{1-\lambda}}{(z-1)(z+\lambda)} \mu(1-\lambda)[-\nabla f(x_k)]_z. \tag{2.35}$$

For NAGD, we have $G(z) = \frac{z+\frac{\lambda}{1-\lambda}}{(z-1)(z+\lambda)}$, again, with an integrator in the forward loop (Fig. 2.8). NAGD is a second-order algorithm with an additional pole at $z = -\lambda$ and a zero at $z = \frac{-\lambda}{1-\lambda}$.

"Can $G(z)$ be of higher order or fractional order"? Of course it can! As shown in Fig. 2.8, a necessary condition for the stability of an algorithm is that all the poles of the closed-loop system are within the unit disc. If the Lipschitz continuous gradient constant L is given, one can replace A with L, and then, the condition is sufficient. For each $G(z)$, there is a corresponding iterative optimization algorithm. $G(z)$ can be a third or higher order system. Apparently, $G(z)$ can also be a fractional-order system. Considering a general second-order discrete system:

$$G(z) = \frac{z+b}{(z-1)(z-a)}, \tag{2.36}$$

the corresponding iterative algorithm is Eq. (2.23). As mentioned earlier, when setting $b = -a/(1+a)$, one can derive the NAGD. When $b = 0$, one can derive the momentum GD. The iterative algorithm can be viewed as a state-space realization of the corresponding system. Thus, it may have many different realizations (all are equivalent). Since two parameters a and b are introduced for a general second-order algorithm design, we used the integral squared error (ISE) as the criterion to optimize the parameters. This is because for different target functions $f(x)$, the Lipschitz continuous gradient constant is different. Thus, the loop forward gain is defined as $\rho := \mu A$.

According to the experimental results (Table 2.3), interestingly, it is found that the optimal a and b satisfy $b = -a/(1+a)$, which is the same design as NAGD. Other criteria such as the IAE and ITAE were used to find other optimal

Table 2.3 General second-order algorithm design. The parameter ρ is the loop forward gain; see text for more details

ρ	0.4	0.8	1.2	1.6	2.0	2.4
a	−0.6	−0.2	0.2	0.6	1	1.4
b	1.5	0.25	−0.1667	−0.3750	−0.5	−0.5833

parameters, but the results are the same as for the ISE. Differently from for NAGD, the parameters were determined by search optimization rather than by mathematical design, which can be extended to more general cases. The algorithms were then tested using the MNIST dataset (Fig. 2.9). It is obvious that for different zeros and poles, the performance of the algorithms is different. One finds that both the $b = -0.25$ and $b = -0.5$ cases perform better than does the SGD momentum. Additionally, both $b = 0.25$ and $b = 0.5$ perform worse. It is also shown that an additional zero can improve the performance, if adjusted properly. It is interesting to observe that both the method and the Nesterov method give an optimal choice of the zero, which is closely related to the pole ($b = -a/(1 + a)$).

Now, let us consider a general third-order discrete system:

$$G(z) = \frac{z^2 + cz + d}{(z-1)(z^2 + az + b)}. \tag{2.37}$$

Set $b = d = 0$; it will reduce to the second-order algorithm discussed above. Compared with the second-order case, the poles can now be complex numbers. More generally, a higher order system can contain more internal models. If all the poles are real, then:

$$G(z) = \frac{1}{(z-1)} \frac{(z-c)}{(z-a)} \frac{(z-d)}{(z-b)}, \tag{2.38}$$

whose corresponding iterative optimization algorithm is

$$\begin{cases} y_{k+1} = y_k - \mu \nabla f(x_k), \\ z_{k+1} = a z_k + y_{k+1} - c y_k, \\ x_{k+1} = b x_k + z_{k+1} - d z_k. \end{cases} \tag{2.39}$$

After some experiments (Table 2.4), it was found that since the ISE was used for tracking a step signal (it is quite simple), the optimal poles and zeros are the same as for the second-order case with a pole-zero cancelation. This is an interesting discovery. In this optimization result, all the poles and zeros are real, and the resulting performance is not very good, as expected. Compare this with the second-order case; the only difference is that in the latter, complex poles can

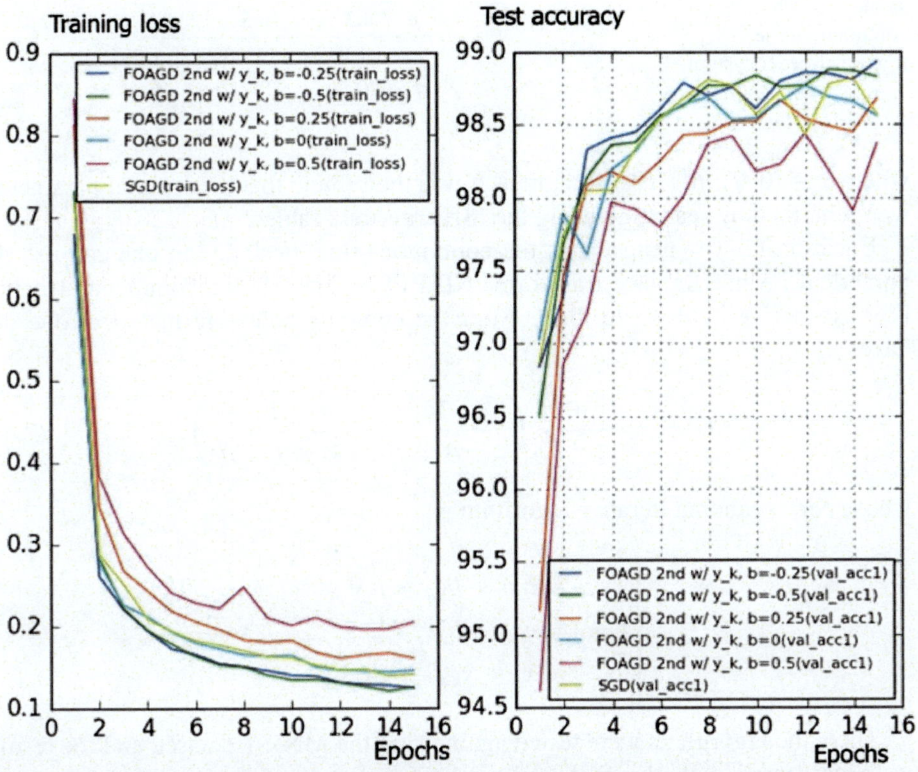

Fig. 2.9 Training loss (**left**); test accuracy (**right**). It is obvious that for different zeros and poles, the performance of the algorithms is different. One finds that both the $b = -0.25$ and $b = -0.5$ cases perform better than does the stochastic gradient descent (SGD) momentum. Additionally, both $b = 0.25$ and $b = 0.5$ perform worse. It is also shown that an additional zero can improve the performance, if adjusted carefully (Courtesy of Professor Yuquan Chen)

Table 2.4 General third-order algorithm design, with parameters defined by Eq. (2.39)

ρ	0.4	0.8	1.2	1.6	2.0	2.4
a	0.6439	0.5247	−0.4097	−0.5955	−1.0364	−1.4629
b	0.0263	0.0649	0.0419	−0.0398	0.0364	0.0880
c	1.5439	0.5747	−0.3763	−0.3705	−0.5364	−0.6462
d	0.0658	0.0812	0.0350	−0.0408	0.0182	0.0367

possibly appear. Thus, the question arises: "How do complex poles play a role in the design?" The answer is obvious: by fractional calculus!

Inspired by M. Jordan's idea in the frequency domain, a continuous-time fractional-order system was designed:

$$G(s) = \frac{1}{s(s^{\alpha} + \beta)}, \tag{2.40}$$

Table 2.5 The continuous-time fractional-order system

ρ	**0.3**	**0.5**	**0.7**	**0.9**
α	1.8494	1.6899	1.5319	1.2284
β	20	20	20	20

where $\alpha \in (0, 2)$, $\beta \in (0, 20]$ at first. It was then found that the optimal parameters were obtained by searching using the ISE criterion (Table 2.5).

Equation (2.40) encapsulates the continuous-time design, and one can use the numerical inverse Laplace transform (NILP) [69] and Matlab command **stmcb( )** [161] to derive its discrete form. After the complex poles are included, one can have:

$$G(z) = \frac{(z+c)}{(z-1)}\left(\frac{1}{z-a+jb} + \frac{1}{z-a-jb}\right), \tag{2.41}$$

whose corresponding iterative algorithm is:

$$\begin{cases} y_{k+1} = ay_k - bz_k - \mu\nabla f(x_k), \\ z_{k+1} = az_k + by_k, \\ x_{k+1} = x_k + y_{k+1} + cy_k. \end{cases} \tag{2.42}$$

Then, the algorithms were tested again using the MNIST dataset, and the results were compared with the SGD's. For the fractional order, $\rho = 0.9$ was used, $a = 0.6786$, $b = 0.1354$, and different values for zero c were used. When $c = 0$, the result was similar to that for the second-order SGD. When c was not equal to 0, the result was similar to that for the second-order NAGD. For the SGD, α was set to be 0.9, and the learning rate was 0.1 (Fig. 2.10). Both $c = 0$ and $c = 0.283$ perform better than the SGD momentum; generally, with appropriate values of c, better performance can be achieved than in the second-order cases. The simulation results demonstrate that fractional calculus (complex poles) can potentially improve the performance, which is closely related to the learning rate.

In general, M. Jordan asked the question: “is there an optimal way to optimize?”. Our answer is a resounding yes, by limiting dynamics analysis and discretization and SGD with other randomness, such as Langevin motion. Herein, the question posed was: “is there a more optimal way to optimize?” Again, the answer is yes, but it requires the fractional calculus to be used to optimize the randomness in SGD, random search and the IMP. There is more potential for further investigations along this line of ideas.

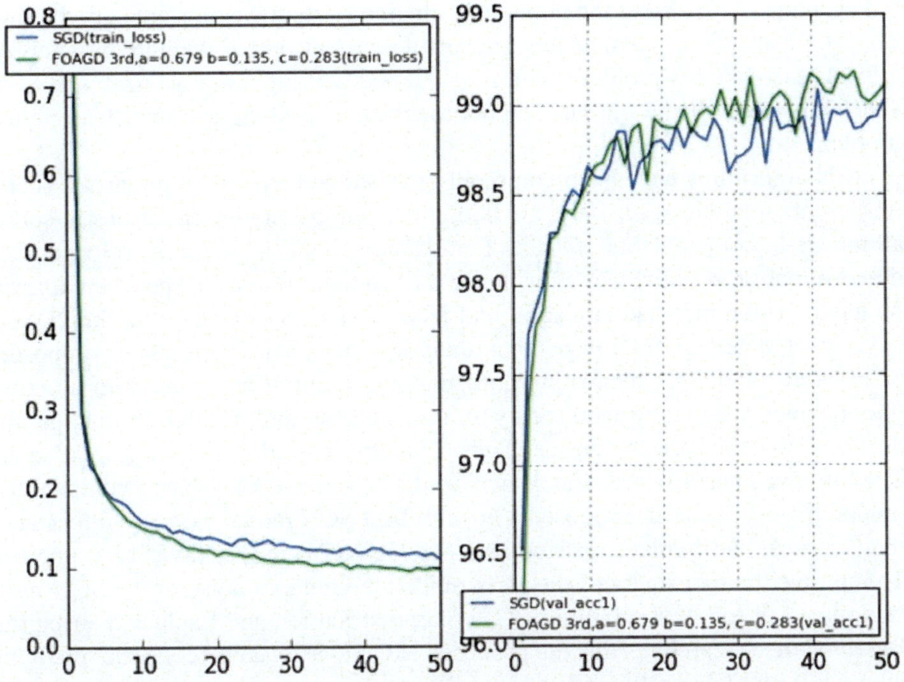

Fig. 2.10 Training loss (**left**); test accuracy (**right**). (Courtesy of Professor Yuquan Chen)

2.7 Case Study: Optimal Randomness for Stochastic Configuration Network (SCN) with Heavy-Tailed Distributions

2.7.1 Introduction

The Stochastic Configuration Network (SCN) model is generated incrementally by using stochastic configuration (SC) algorithms [145]. Compared with the existing randomized learning algorithms for single-layer feed-forward neural networks (SLFNNs), the SCN can randomly assign the input weights (w) and biases (b) of the hidden nodes in a supervisory mechanism, which is selecting the random parameters with an inequality constraint and assigning the scope of the random parameters adaptively. It can ensure that the built randomized learner models have universal approximation property. Then, the output weights are analytically evaluated in either a constructive or selective manner [145]. In contrast with the known randomized learning algorithms, such as the Randomized Radial Basis Function (RBF) Networks [14] and the Random Vector Functional-link (RVFL) [108], SCN can provide good generalization performance at a faster speed. Concretely, there are three types of SCN algorithms, which are SC-I, SC-II, and SC-III. SC-I algorithm uses a constructive scheme to evaluate the output weights only for the newly added hidden node [146]. All of the previously obtained output weights are kept the same. The SC-II algorithm recalculates part of the current output weights by analyzing a local least-squares problem with user-defined shifting window size. The SC-III algorithm finds all the output weights together by solving a global least-squares problem.

SCN algorithms have been commonly studied and used in many areas, such as image data analytics [73, 101], prediction of component concentrations in sodium aluminate liquor [147], etc. [58, 82]. For example, in [73], Li et al. developed a two-dimensional SCNs (2DSCNs) for image data modeling tasks. Experimental results on hand written digit classification and face recognition showed that the 2DSCNs have great potential for image data analytics. In [147], Wang et al. proposed a SCN-based model for measuring component concentrations in sodium aluminate liquor, which are usually acquired by titration analysis and suffered from larger time delays. From the results, the mechanism model showed the internal relationship. The improved performance can be achieved by using the SCN-based compensation model. In [81], Lu et al. proposed a novel robust SCN model based on a mixture of the Gaussian and Laplace distributions (MoGL-SCN) in the Bayesian framework. To improve the robustness of the SCN model, the random noise of the SCN model is assumed to follow a mixture of Gaussian distribution and Laplace distributions. Based on the research results, the proposed MOGL-SCN could construct prediction intervals with higher reliability and prediction accuracy.

Neural Networks (NNs) can learn from data to train feature-based predictive models. However, the learning process can be time-consuming and infeasible for applications with data streams. An optimal method is to randomly assign the weights of the NNs so that the task can become a linear least-squares problem. In [122], Wang et al. classified the NN models into three types: first, the feed-forward networks with random weights (RW-FNN) [107], second, recurrent NNs with random weights [86], third, randomized kernel approximations [70]. According to [122], there are three benefits of the randomness: (1) simplicity of implementation, (2) faster learning and less human intervention, (3) possibility of leveraging linear regression and classification algorithms. Randomness is used to define a feature map, which converts the data input into a high dimensional space where learning is more simpler. The resulting optimization problem becomes a standard linear least-squares, which is a simpler and scalable learning procedure.

For the original SCN algorithms, weights and biases are randomly generated in uniform distribution. Randomness plays a significant role in both exploration and exploitation. A good NNs architecture with randomly assigned weights can easily outperform a more deficient architecture with finely tuned weights [122]. Therefore, it is critical to discuss the optimal randomness for the weights and biases in SCN algorithms. In this study, the authors mainly discussed the impact of three different heavy-tailed distributions on the performance of the SCN algorithms, Lévy distribution, Cauchy distribution, and Weibull distribution [116]. Heavy-tailed distribution has shown optimal randomness for finding targets [149], which plays a significant role in exploration and exploitation [151]. It is important to point out that the proposed SCN models are very different from Lu et al. [81]. As mentioned earlier, Lu at al. assumed that the random noise of the SCN model following a mixture of Gaussian distribution and Laplace distributions. In this research study, the author randomly initializes the weights and biases with heavy-tailed distributions instead of uniform distribution. To compare with the mixture distributions, the author also used the mixture distributions for weight and bias generation. A more detailed comparison of the two heavy-tailed methods is shown in Results and Discussion section.

There are two objectives for this research: (1) compare the performance of SCN algorithms with heavy-tailed distributions on a linear regression model [138]; (2) evaluate the SCN algorithms performance on MNIST handwritten digit classification problem with heavy-tailed distributions. The rest of this section is organized as follows: Sect. 2.2 introduces fundamental definitions of the heavy-tailed distributions and how to generate the random numbers according to the given distribution. Results and discussion are presented in Sect. 2.3. A simple regression model and MNIST handwritten digit classification problem are used to demonstrate the heavy-tailed SCNs. In Sect. 2.4, the author draw conclusive remarks and share views in SCN with heavy-tailed distributions in future research.

2.7.2 SCN with Heavy-Tailed PDFs

For the original SCN algorithms, weights and biases are randomly generated using a uniform PDF. Randomness plays a significant role in both exploration and exploitation. A good neural network architecture with randomly assigned weights can easily outperform a more deficient architecture with finely tuned weights [122]. Therefore, it is critical to discuss the optimal randomness for the weights and biases in SCN algorithms. Heavy-tailed PDFs have shown optimal randomness for finding targets [19, 149], which plays a significant role in exploration and exploitation [151]. Therefore, herein, heavy-tailed PDFs were used to randomly update the weights and biases in the hidden layers to determine if the SCN models display improved performance. Some of the key parameters of the SCN models are listed in Table 2.6. For example, the maximum times of random configuration T_{max} are set as 200. The scale factor lambda in the activation function, which directly determines the range for the random parameters, was examined by using different settings (0.5–200). The tolerance was set as 0.05. Most of the parameters for the SCN with heavy-tailed PDFs were kept the same with the original SCN algorithms for comparison purposes. For more details, please refer to [145].

2.7.3 A Regression Model and Parameter Tuning

The dataset of the regression model was generated by a real-valued function [138]:

Table 2.6 SCNs with key parameters

Properties	Values
Name:	"Stochastic Configuration Networks"
version:	"1.0 beta"
L:	hidden node number
W:	input weight matrix
b:	hidden layer bias vector
Beta:	output weight vector
r:	regularization parameter
tol:	tolerance
Lambdas:	random weights range
L_{max}:	maximum number of hidden neurons
T_{max}:	maximum times of random configurations
nB:	number of node being added in one loop

$$f(x) = 0.2e^{-(10x-4)^2} + 0.5e^{-(80x-40)^2} + 0.3e^{-(80x-20)^2}, \tag{2.43}$$

where x ∈ [0, 1]. There were 1000 points randomly generated from the uniform distribution on the unit interval [0, 1] in the training dataset. The test set had 300 points generated from a regularly spaced grid on [0, 1]. The input and output attributes were normalized into [0, 1], and all the results reported in this research represent averages over 1000 independent trials. The settings of the parameters were similar to for the SCN in [145].

Heavy-tailed PDF algorithms have user-defined parameters, for example, the power law index for SCN-Lévy, and location and scale parameters for SCN-Cauchy and SCN-Weibull, respectively. Thus, to illustrate the effect of parameters on the optimization results and to offer reference values for the proposed SCN algorithms, parameter analysis was conducted, and corresponding experiments were performed. Based on the experimental results, for the SCN-Lévy algorithm, the most optimal power law index is 1.1 for achieving the minimum number of hidden nodes. For the SCN-Weibull algorithm, the optimal location parameter α and scale parameter β for the minimum number of hidden nodes are 1.9 and 0.2, respectively. For the SCN-Cauchy algorithm, the optimal location parameter α and scale parameter β for the minimum number of hidden nodes are 0.9 and 0.1, respectively.

2.7.3.1 Performance Comparison Among SCNs with Heavy-Tailed PDFs

In Table 2.7, the performance of SCN, SCN-Lévy, SCN-Cauchy, SCN-Weibull, and SCN-Mixture are shown, in which mean values are reported based on 1000 independent trials. Wang et al. [145] used time cost to evaluate the SCN algorithms' performance. In the present study, the author used the mean hidden node numbers to evaluate the performance. The number of hidden nodes is associated with modeling

Table 2.7 Performance comparison of SCN models on regression problem

Models	Mean hidden node number	RMSE
SCN	75 ± 5	0.0025,
SCN-Lévy	70 ± 6	0.0010,
SCN-Cauchy	59 ± 3	0.0057,
SCN-Weibull	63 ± 4	0.0037,
SCN-Mixture	70 ± 5	0.0020.

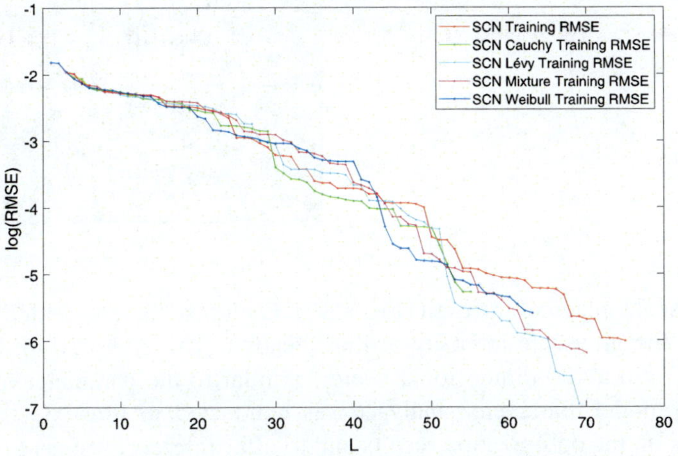

Fig. 2.11 Performance of SCN, SCN-Lévy, SCN-Weibull, SCN-Cauchy and SCN-Mixture. The parameter L is the hidden node number

accuracy. Therefore, herein, the analysis determined if an SCN with heavy-tailed PDFs used fewer hidden nodes to generate high performance, which would make the NNs less complex. According to the numerical results, the SCN-Cauchy used the lowest number of mean hidden nodes, 59, with an root mean squared error (RMSE) of 0.0057. The SCN-Weibull had a mean number of 63 hidden nodes, with an RMSE of 0.0037. The SCN-Mixture had a mean number of 70 hidden nodes, with an RMSE of 0.0020. The mean number of hidden nodes for SCN-Lévy was also 70. The original SCN model had a mean number of 75 hidden nodes. A more detailed training process is shown in Fig. 2.11. With fewer hidden node numbers, the SCN models with heavy-tailed PDFs can be faster than the original SCN model. The neural network structure is also less complicated than the SCN. Our numerical results for the regression task demonstrate remarkable improvements in modeling performance compared with the current SCN model results.

2.7.4 MNIST Handwritten Digit Classification

The handwritten digit dataset contains 4000 training examples and 1000 testing examples, a subset of the MNIST handwritten digit dataset. Each image is a 20×20

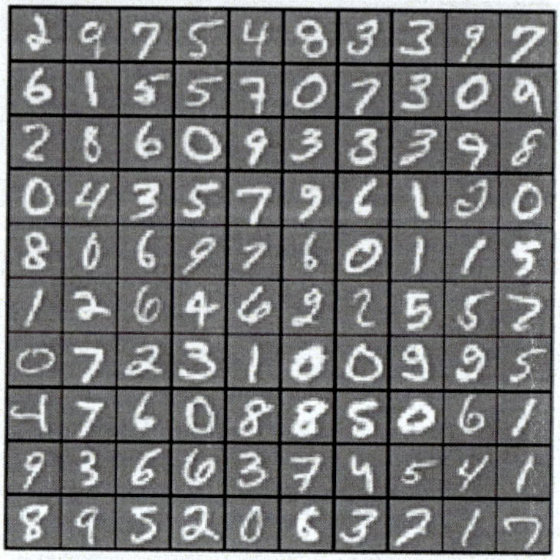

Fig. 2.12 The handwritten digit dataset example

pixel grayscale image of the digit (Fig. 2.12). Each pixel is represented by a number indicating the grayscale intensity at that location. The 20-by-20 grid of pixels is "unrolled" into a 400-dimensional vector. Similar to the parameter tuning for the regression model, parameter analysis was conducted to illustrate the impact of parameters on the optimization results and to offer reference values for the MNIST handwritten digit classification SCN algorithms. Corresponding experiments were performed. According to the experimental results, for the SCN-Lévy algorithm, the most optimal power law index is 1.6 for achieving the best RMSE performance. For the SCN-Cauchy algorithm, the optimal location parameter α and scale parameter β for the lowest RMSE are 0.2 and 0.3, respectively.

2.7.4.1 Performance Comparison Among SCNs on MNIST

The performance of the SCN, SCN-Lévy, SCN-Cauchy, and SCN-Mixture are shown in Table 2.8. Based on the experimental results, the SCN-Cauchy, SCN-Lévy and SCN-Mixture have better performance in training and test accuracy, compared with the original SCN model. A detailed training process is shown in Fig. 2.13. Within around 100 hidden nodes, the SCN models with heavy-tailed PDFs perform similarly to the original SCN model. When the number of the hidden nodes is greater than 100, the SCN models with heavy-tailed PDFs have lower RMSEs. Since more parameters for weights and biases are initialized in heavy-tailed PDFs, this may cause an SCN with heavy-tailed PDFs to converge to the optimal values at a faster speed. The experimental results for the MNIST handwritten classification problem demonstrate improvements in modeling performance. They also show that SCN models with heavy-tailed PDFs have a better search ability for achieving lower RMSEs.

Table 2.8 Performance comparison between SCN, SCN-Lévy and SCN-Cauchy

Models	Training accuracy	Test accuracy
SCN	$94.0 \pm 1.9\%$	$91.2 \pm 6.2\%$,
SCN-Lévy	$94.9 \pm 0.8\%$	$91.7 \pm 4.5\%$,
SCN-Cauchy	$95.4 \pm 1.3\%$	$92.4 \pm 5.5\%$.
SCN-Mixture	$94.7 \pm 1.1\%$	$91.5 \pm 5.3\%$.

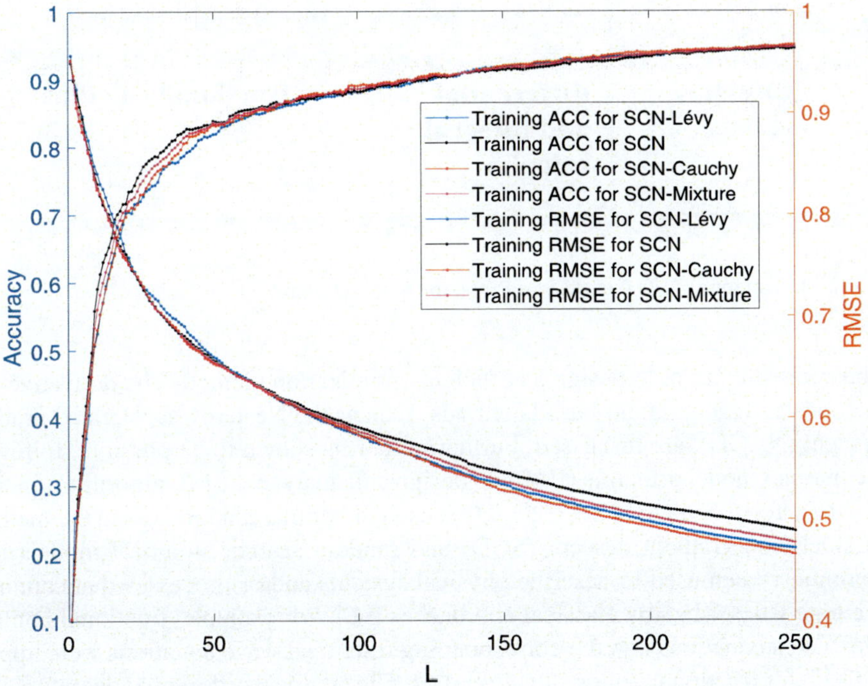

Fig. 2.13 Classification performance of SCNs

2.8 Chapter Summary

Big data and machine learning (ML) are two of the hottest topics of applied scientific research, and they are closely related to one another. To better understand them, in this chapter, the authors advocate fractional calculus (FC), as well as fractional-order thinking (FOT), for big data and ML analysis and applications. In Sect. 2.4, we discussed the relationships between big data, variability and FC, as well as why fractional-order data analytics (FODA) should be used and what it is. The topics included the Hurst parameter, fractional Gaussian noise (fGn), fractional Brownian motion (fBm), fractional autoregressive integrated moving average (FARIMA), formalism of continuous-time random walk (CTRW), unmanned aerial vehicles (UAVs), and digital agriculture (PA).

In Sect. 2.5, how to learn efficiently (optimally) for ML algorithms is discussed. The key to developing an efficient learning process is the method of optimization.

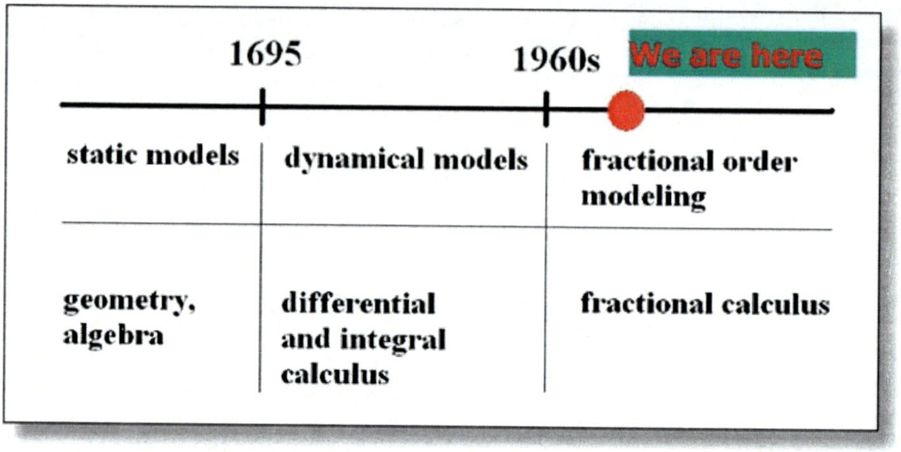

Fig. 2.14 Timeline of FC (courtesy of Professor Igor Podlubny)

Thus, it is important to design an efficient optimization method. The derivative-free methods and the gradient-based methods, such as the Nesterov accelerated gradient descent (NAGD), are discussed. Furthermore, it is shown to be possible, following the internal model principle (IMP), to design and analyze the ML algorithms in the S or Z transform domain in Sect. 2.6. FC is used in optimal randomness in the methods of stochastic gradient descent (SGD) and random search. Nonlocal models have commonly been used to describe physical systems and/or processes that cannot be accurately described by classical approaches [96]. For example, fractional nonlocal Maxwell's equations and the corresponding fractional wave equations were applied in [135] for fractional vector calculus [105]. The nonlocal differential operators [34], including nonlocal analogs of the gradient/Hessian, are the key of these nonlocal models, which could lead to very interesting research with FC in the near future.

Fractional dynamics is a response to the need for a more advanced characterization of our complex world to capture structure at very small or very large scales that had previously been smoothed over. If one wishes to obtain results that are better than the best possible using integer-order calculus-based methods, or are "more optimal," we advocate applying FOT and going fractional! In this era of big data, decision and control need FC, such as fractional-order signals, systems, and controls. The future of ML should be physics-informed, scientific (cause–effect embedded or cause–effect discovery) and involving the use of FC, where the modeling is closer to nature. Laozi (unknown, around the sixth century to fourth century BC), the ancient Chinese philosopher, is said to have written a short book *Dao De Jing*(*Tao Te Ching*), in which he observed: "The Tao that can be told is not the eternal Tao" [49]. People over thousands of years have shared different understandings of the meaning of the Tao. Our best understanding of the Tao is nature, whose rules of complexity can be explained in a non-normal way. Fractional dynamics, FC, and heavy-tailedness may well be that non-normal way (Fig. 2.14), at least for the not-too-distant future.

References

1. Abel, N.: Solution of a couple of problems by means of definite integrals. Magazin for Naturvidenskaberne **2**(55), 2 (1823)
2. Ahn, S., Kim, J.H., Ramaswami, V.: A new class of models for heavy tailed distributions in finance and insurance risk. Insur. Math. Econ. **51**(1), 43–52 (2012)
3. An, W., Wang, H., Sun, Q., Xu, J., Dai, Q., Zhang, L.: A PID controller approach for stochastic optimization of deep networks. In: Proceedings of the IEEE Conference on Computer Vision and Pattern Recognition, pp. 8522–8531 (2018)
4. Arabas, J., Opara, K.: Population diversity of non-elitist evolutionary algorithms in the exploration phase. IEEE Trans. Evol. Comput. **24**(6), 1050–1062 (2019)
5. Asmussen, S.: Steady-state properties of $GI/G/1$. In: Applied Probability and Queues, pp. 266–301 (2003)
6. Bahat, D., Rabinovitch, A., Frid, V.: Tensile Fracturing in Rocks. Springer, Berlin (2005)
7. Bahg, G., Evans, D.G., Galdo, M., Turner, B.M.: Gaussian process linking functions for mind, brain, and behavior. Proc. Natl. Acad. Sci. **117**(47), 29398–29406 (2020)
8. Barabási, A.L., Albert, R.: Emergence of scaling in random networks. Science **286**(5439), 509–512 (1999)
9. Bardi, J.S.: The Calculus Wars: Newton, Leibniz, and the Greatest Mathematical Clash of All Time. Hachette UK (2009)
10. Bernardi, M., Petrella, L.: Interconnected risk contributions: a heavy-tail approach to analyze US financial sectors. J. Risk Financ. Manag. **8**(2), 198–226 (2015)
11. Boutahar, M., Dufrénot, G., Péguin-Feissolle, A.: A simple fractionally integrated model with a time-varying long memory parameter d_t. Comput. Econ. **31**(3), 225–241 (2008)
12. Boyer, C.B.: The History of the Calculus and its Conceptual Development: (The Concepts of the Calculus). Courier Corporation, North Chelmsford (1959)
13. Brockwell, P.J., Davis, R.A., Fienberg, S.E.: Time Series: Theory and Methods. Springer, New York (1991)
14. Broomhead, D., Lowe, D.: Multivariable functional interpolation and adaptive networks. Complex Syst. **2**, 321–355 (1988)
15. Burnecki, K., Weron, A.: Lévy stable processes. From stationary to self-similar dynamics and back. An application to finance. Acta Physica Polonica Series B **35**(4), 1343–1358 (2004)
16. Che, Y., Wang, Q., Xie, Z., Zhou, L., Li, S., Hui, F., Wang, X., Li, B., Ma, Y.: Estimation of maize plant height and leaf area index dynamic using unmanned aerial vehicle with oblique and nadir photography. Ann. Bot. **126**(4), 765–773 (2020)
17. Chen, D., Sun, S., Zhang, C., Chen, Y., Xue, D.: Fractional-order TV-L 2 model for image denoising. Centr. Eur. J. Phys. **11**(10), 1414–1422 (2013)
18. Chen, D., Xue, D., Chen, Y.: More optimal image processing by fractional order differentiation and fractional order partial differential equations. In: Proceedings of the International Symposium on Fractional PDEs (2013)
19. Chen, Y.: Fundamental principles for fractional order gradient methods. Ph.D. thesis, University of Science and Technology of China, China (2020)
20. Chen, Y., Sun, R., Zhou, A.: An overview of fractional order signal processing (FOSP) techniques. In: Proceedings of the International Design Engineering Technical Conferences and Computers and Information in Engineering Conference (2007)
21. Chen, Y., Sun, R., Zhou, A.: An improved Hurst parameter estimator based on fractional Fourier transform. Telecommun. Syst. **43**(3-4), 197–206 (2010)
22. Clegg, R.G.: A practical guide to measuring the Hurst parameter. arXiv preprint math/0610756 (2006)
23. Cottone, G., Di Paola, M.: On the use of fractional calculus for the probabilistic characterization of random variables. Probab. Eng. Mech. **24**(3), 321–330 (2009)
24. Cottone, G., Di Paola, M., Metzler, R.: Fractional calculus approach to the statistical characterization of random variables and vectors. Physica A: Stat. Mech. Appl. **389**(5), 909–920 (2010)

25. Crovella, M.E., Bestavros, A.: Self-similarity in World Wide Web traffic: evidence and possible causes. IEEE/ACM Trans. Networking **5**(6), 835–846 (1997)
26. Csete, M., Doyle, J.: Bow ties, metabolism and disease. Trends Biotechnol. **22**(9), 446–450 (2004)
27. Decreusefond, L.: Stochastic analysis of the fractional Brownian motion. Potential Anal. **10**(2), 177–214 (1999)
28. Deng, R., Jiang, Y., Tao, M., Huang, X., Bangura, K., Liu, C., Lin, J., Qi, L.: Deep learning-based automatic detection of productive tillers in rice. Comput. Electron. Agric. **177**, 105703 (2020)
29. Díaz-Varela, R., de la Rosa, R., León, L., Zarco-Tejada, P.: High-resolution airborne UAV imagery to assess olive tree crown parameters using 3D photo reconstruction: Application in breeding trials. Remote Sens. **7**(4), 4213–4232 (2015)
30. Doyle, J.: Universal laws and architectures. In: CDS 212 Lect. Notes (2011)
31. Doyle, J.C., Csete, M.: Architecture, constraints, and behavior. Proc. Natl. Acad. Sci. **108**(Supplement 3), 15624–15630 (2011)
32. Duchi, J., Hazan, E., Singer, Y.: Adaptive subgradient methods for online learning and stochastic optimization. J. Mach. Learn. Res. **12**(7), 2121–2159 (2011)
33. Fan, Y., Koellermeier, J.: Accelerating the convergence of the moment method for the Boltzmann equation using filters. J. Sci. Comput. **84**(1), 1–28 (2020)
34. Feliu Faba, J., Fan, Y., Ying, L.: Meta-learning pseudo-differential operators with deep neural networks. J. Comput. Phys. **408**, 109309 (2020)
35. Feller, W.: An Introduction to Probability Theory and its Application, vol II. Wiley, New York (1971)
36. Feynman, R.P.: The principle of least action in quantum mechanics. In: Feynman's Thesis—A New Approach to Quantum Theory, pp. 1–69. World Scientific, New York (2005)
37. Firican, G.: The 10 Vs of Big Data (2017). https://tdwi.org/articles/2017/02/08/10-vs-of-big-data.aspx
38. Foss, S., Korshunov, D., Zachary, S.: An Introduction to Heavy-tailed and Subexponential Distributions, vol. 6. Springer, New York (2011)
39. Francis, B.A., Wonham, W.M.: The internal model principle of control theory. Automatica **12**(5), 457–465 (1976)
40. Geerolf, F.: A theory of Pareto distributions. UCLA Manuscript (2016)
41. Géron, A.: Hands-on Machine Learning with Scikit-Learn, Keras, and TensorFlow: Concepts, Tools, and Techniques to Build Intelligent Systems. O'Reilly Media, California (2019)
42. Geweke, J., Porter-Hudak, S.: The estimation and application of long memory time series models. J. Time Ser. Anal. **4**(4), 221–238 (1983)
43. Gonzalez-Dugo, V., Goldhamer, D., Zarco-Tejada, P.J., Fereres, E.: Improving the precision of irrigation in a pistachio farm using an unmanned airborne thermal system. Irrig. Sci. **33**(1), 43–52 (2015)
44. Gorenflo, R., Mainardi, F.: Fractional calculus and stable probability distributions. Arch. Mech. **50**(3), 377–388 (1998)
45. Gorenflo, R., Mainardi, F., Vivoli, A.: Continuous-time random walk and parametric subordination in fractional diffusion. Chaos, Solitons Fractals **34**(1), 87–103 (2007)
46. Graves, T., Gramacy, R., Watkins, N., Franzke, C.: A brief history of long memory: Hurst, mandelbrot and the road to ARFIMA, 1951–1980. Entropy **19**(9), 437 (2017)
47. Gray, H.L., Zhang, N.F., Woodward, W.A.: On generalized fractional processes. J. Time Ser. Anal. **10**(3), 233–257 (1989)
48. Gubner, J.A.: Probability and Random Processes for Electrical and Computer Engineers. Cambridge University, Cambridge (2006)
49. Hall, D.L.: Dao De Jing: A Philosophical Translation. Random House Digital, Inc., New York (2003)
50. Hamilton, S.W.R.: On A General Method in Dynamics. Richard Taylor, New York (1834)
51. Hariya, Y., Kurihara, T., Shindo, T., Jin'no, K.: Lévy flight PSO. In: Proceedings of the IEEE Congress on Evolutionary Computation (CEC) (2015)

52. Harmantzis, F.: Heavy network traffic modeling and simulation using stable FARIMA processes. In: Proceedings of the 19th International Teletraffic Congress (ITC19) (2005)
53. Hartley, T.T., Lorenzo, C.F.: Fractional-order system identification based on continuous order-distributions. Signal Process. **83**(11), 2287–2300 (2003)
54. Haubold, H.J., Mathai, A.M., Saxena, R.K.: Mittag-Leffler functions and their applications. J. Appl. Math. **2011** (2011). https://doi.org/10.1155/2011/298628
55. Hawking, S.W.: The path-integral approach to quantum gravity. In: General Relativity: An Einstein centenary survey, pp. 746–789. University Press, United Kingdom (1979)
56. Hilbert, M., López, P.: The world's technological capacity to store, communicate, and compute information. Science **332**(6025), 60–65 (2011)
57. Hilfer, R., Anton, L.: Fractional master equations and fractal time random walks. Phys. Rev. E **51**(2), R848 (1995)
58. Huang, C., Huang, Q., Wang, D.: Stochastic configuration networks based adaptive storage replica management for power big data processing. IEEE Trans. Industr. Inform. **16**(1), 373–383 (2019)
59. Hutton, E.L.: Xunzi: The Complete Text. Princeton University, Princeton (2014)
60. Jayakumar, K.: Mittag-Leffler process. Math. Comput. Model. **37**(12-13), 1427–1434 (2003)
61. Johnson, N.L., Kotz, S., Balakrishnan, N.: Continuous Univariate Distributions. Wiley, New York (1995)
62. Kashima, K., Yamamoto, Y.: System theory for numerical analysis. Automatica **43**(7), 1156–1164 (2007)
63. Kello, C.T., Brown, G.D., Ferrer Cancho, R., Holden, J.G., Linkenkaer Hansen, K., Rhodes, T., Van Orden, G.C.: Scaling laws in cognitive sciences. Trends Cogn. Sci. **14**(5), 223–232 (2010)
64. Kerrigan, E.: What the machine should learn about models for control (2020). https://www.ifac2020.org/program/workshops/machine-learning-meets-model-based-control
65. Kingma, D.P., Ba, J.: Adam: a method for stochastic optimization. arXiv preprint arXiv:1412.6980 (2014)
66. Klafter, J., Lim, S., Metzler, R.: Fractional Dynamics: Recent Advances. World Scientific, Singapore (2012)
67. Ko, M., Stark, B., Barbadillo, M., Chen, Y.: An evaluation of three approaches using Hurst estimation to differentiate between normal and abnormal HRV. In: Proceedings of the International Design Engineering Technical Conferences and Computers and Information in Engineering Conference. American Society of Mechanical Engineers (2015)
68. Koutsoyiannis, D.: The Hurst phenomenon and fractional Gaussian noise made easy. Hydrol. Sci. J. **47**(4), 573–595 (2002)
69. Kuhlman, K.L.: Review of inverse Laplace transform algorithms for Laplace-space numerical approaches. Numer. Algorithms **63**(2), 339–355 (2013)
70. Lee, D.D., Pham, P., Largman, Y., Ng, A.: Advances in neural information processing systems. In: Proceedings of the 26th Annual Conference on Neural Information Processing Systems 2012, NIPS 2012 (2009)
71. Lévy, M., Solomon, S.: New evidence for the power-law distribution of wealth. Physica A: Stat. Mech. Appl. **242**(1-2), 90–94 (1997)
72. Li, M.: Modeling autocorrelation functions of long-range dependent teletraffic series based on optimal approximation in Hilbert space—A further study. Appl. Math. Model. **31**(3), 625–631 (2007)
73. Li, M., Wang, D.: 2-D stochastic configuration networks for image data analytics. IEEE Trans. Cybern. **51**(1), 359–372 (2019)
74. Li, N., Cruz, J., Chien, C.S., Sojoudi, S., Recht, B., Stone, D., Csete, M., Bahmiller, D., Doyle, J.C.: Robust efficiency and actuator saturation explain healthy heart rate control and variability. Proc. Natl. Acad. Sci. **111**(33), E3476–E3485 (2014)
75. Li, Q., Tricaud, C., Sun, R., Chen, Y.: Great Salt Lake surface level forecasting using FIGARCH model. In: Proceedings of the International Design Engineering Technical Conferences and Computers and Information in Engineering Conference, vol. 4806, pp. 1361–1370 (2007)

76. Li, Z., Liu, L., Dehghan, S., Chen, Y., Xue, D.: A review and evaluation of numerical tools for fractional calculus and fractional order controls. Int. J. Control. **90**(6), 1165–1181 (2017)
77. Liakos, K.G., Busato, P., Moshou, D., Pearson, S., Bochtis, D.: Machine learning in agriculture: A review. Sensors **18**(8), 2674 (2018)
78. Liu, K., Chen, Y., Zhang, X.: An evaluation of ARFIMA (autoregressive fractional integral moving average) programs. Axioms **6**(2), 16 (2017)
79. Liu, K., Domański, P.D., Chen, Y.: Control performance assessment with fractional lower order moments. In: Proceedings of the 7th International Conference on Control, Decision and Information Technologies (CoDIT), vol. 1, pp. 778–783. IEEE, New York (2020)
80. Liu, T., Zhang, P., Dai, W.S., Xie, M.: An intermediate distribution between Gaussian and Cauchy distributions. Physica A: Stat. Mech. Appl. **391**(22), 5411–5421 (2012)
81. Lu, J., Ding, J.: Mixed-distribution-based robust stochastic configuration networks for prediction interval construction. IEEE Trans. Industr. Inform. **16**(8), 5099–5109 (2019)
82. Lu, J., Ding, J., Dai, X., Chai, T.: Ensemble stochastic configuration networks for estimating prediction intervals: A simultaneous robust training algorithm and its application. IEEE Trans. Neural Networks Learn. Syst. **31**(12), 5426–5440 (2020)
83. Luchko, Y., Mainardi, F.: Some properties of the fundamental solution to the signalling problem for the fractional diffusion-wave equation. Open Phys. **11**(6), 666–675 (2013)
84. Luchko, Y., Mainardi, F.: Cauchy and signaling problems for the time-fractional diffusion-wave equation. J. Vib. Acoust. **136**(5), 050904 (2014)
85. Luchko, Y., Mainardi, F., Povstenko, Y.: Propagation speed of the maximum of the fundamental solution to the fractional diffusion–wave equation. Comput. Math. Appl. **66**(5), 774–784 (2013)
86. Lukoševičius, M., Jaeger, H.: Reservoir computing approaches to recurrent neural network training. Comput. Sci. Rev. **3**(3), 127–149 (2009)
87. Ma, X., Nikias, C.L.: Joint estimation of time delay and frequency delay in impulsive noise using fractional lower order statistics. IEEE Trans. Signal Process. **44**(11), 2669–2687 (1996)
88. Mainardi, F.: The fundamental solutions for the fractional diffusion-wave equation. Appl. Math. Lett. **9**(6), 23–28 (1996)
89. Mandelbrot, B.: The Pareto-Lévy law and the distribution of income. Int. Econ. Rev. **1**(2), 79–106 (1960)
90. Mandelbrot, B.B., Van Ness, J.W.: Fractional Brownian motions, fractional noises and applications. SIAM Rev. **10**(4), 422–437 (1968)
91. Mandelbrot, B.B., Wallis, J.R.: Robustness of the rescaled range r/s in the measurement of noncyclic long run statistical dependence. Water Resour. Res. **5**(5), 967–988 (1969)
92. Metzler, R., Klafter, J.: The random walk's guide to anomalous diffusion: a fractional dynamics approach. Phys. Rep. **339**(1), 1–77 (2000)
93. Mitchell, T.M.: Machine Learning. McGraw hill, New York (1997)
94. Montroll, E.W., Weiss, G.H.: Random walks on lattices. II. J. Math. Phys. **6**(2), 167–181 (1965)
95. Montroll, E.W., West, B.J.: On an enriched collection of stochastic processes. Fluctuation Phenomena **66**, 61 (1979)
96. Nagaraj, S.: Optimization and learning with nonlocal calculus. arXiv preprint arXiv:2012.07013 (2020)
97. Nakahira, Y., Liu, Q., Sejnowski, T.J., Doyle, J.C.: Diversity-enabled sweet spots in layered architectures and speed-accuracy trade-offs in sensorimotor control. arXiv preprint arXiv:1909.08601 (2019)
98. Nesterov, Y.: A method for unconstrained convex minimization problem with the rate of convergence $O(1/k^2)$. Dokl. Akad. Nauk Russ. Acad. Sci. **269**, 543–547 (1983)
99. Niu, H., Chen, Y., Chen, Y.: Fractional-order extreme learning machine with Mittag-Leffler distribution. In: International Design Engineering Technical Conferences and Computers and Information in Engineering Conference. American Society of Mechanical Engineers (2019)
100. Niu, H., Hollenbeck, D., Zhao, T., Wang, D., Chen, Y.: Evapotranspiration estimation with small UAVs in precision agriculture. Sensors **20**(22), 6427 (2020)

101. Niu, H., Wang, D., Chen, Y.: Estimating actual crop evapotranspiration using deep stochastic configuration networks model and UAV-based crop coefficients in a pomegranate orchard. In: Autonomous Air and Ground Sensing Systems for Agricultural Optimization and Phenotyping V. International Socicty for Optics and Photonics (2020)
102. Niu, H., Wang, D., Chen, Y.: Estimating crop coefficients using linear and deep stochastic configuration networks models and UAV-based normalized difference vegetation index (NDVI). In: Proceedings of the 2020 International Conference on Unmanned Aircraft Systems (ICUAS), pp. 1485–1490. IEEE, New York (2020)
103. Niu, H., Zhao, T., Wang, D., Chen, Y.: Estimating evapotranspiration with UAVs in agriculture: A review. In: Proceedings of the ASABE Annual International Meeting. American Society of Agricultural and Biological Engineers (2019)
104. Niu, H., Zhao, T., Wang, D., Chen, Y.: A UAV resolution and waveband aware path planning for onion irrigation treatments inference. In: 2019 International Conference on Unmanned Aircraft Systems (ICUAS), pp. 808–812. IEEE, New York (2019)
105. Ortigueira, M., Machado, J.: On fractional vectorial calculus. Bull. Pol. Acad. Sci. Tech. Sci. **66**(4), 389–402 (2018)
106. Ortigueira, M.D., Batista, A.G.: On the relation between the fractional Brownian motion and the fractional derivatives. Phys. Lett. A **372**(7), 958–968 (2008)
107. Pao, Y.H., Park, G.H., Sobajic, D.J.: Learning and generalization characteristics of the random vector functional-link net. Neurocomputing **6**(2), 163–180 (1994)
108. Pao, Y.H., Takefuji, Y.: Functional-link net computing: theory, system architecture, and functionalities. Computer **25**(5), 76–79 (1992)
109. Pesquet-Popescu, B., Pesquet, J.C.: Synthesis of bidimensional α-stable models with long-range dependence. Signal Process. **82**(12), 1927–1940 (2002)
110. Pipiras, V., Taqqu, M.S.: Long-range Dependence and Self-similarity, vol. 45. Cambridge University, Cambridge (2017)
111. Podlubny, I., Magin, R.L., Trymorush, I.: Niels Henrik Abel and the birth of fractional calculus. Fract. Calc. Appl. Anal. **20**(5), 1068–1075 (2017)
112. Polyak, B.T.: Some methods of speeding up the convergence of iteration methods. USSR Comput. Math. Math. Phys. **4**(5), 1–17 (1964)
113. Pramukkul, P., Svenkeson, A., Grigolini, P., Bologna, M., West, B.: Complexity and the fractional calculus. Adv. Math. Phys. **2013**, 1–7 (2013)
114. Reinsel, D., Gantz, J., Rydning, J.: Data age 2025: the evolution of data to life-critical don't focus on big data; focus on the data that's big. In: International Data Corporation (IDC) White Paper (2017)
115. Resnick, S.I.: Heavy-tail Phenomena: Probabilistic and Statistical Modeling. Springer Science & Business Media, New York (2007)
116. Rinne, H.: The Weibull Distribution: A Handbook. CRC Press, New York (2008)
117. Rolski, T., Schmidli, H., Schmidt, V., Teugels, J.L.: Stochastic Processes for Insurance and Finance, vol. 505. Wiley, New York (2009)
118. RongHua, F.: Modeling and application of theory based on time series ARMA. Sci. Tech. Inf. **2012**(19), 153 (2012)
119. Ross, B.: The development of fractional calculus 1695–1900. Hist. Math. **4**(1), 75–89 (1977)
120. Samoradnitsky, G.: Stable non-Gaussian Random Processes: Stochastic Models with Infinite Variance. Routledge, England (2017)
121. Samorodnitsky, G.: Long range dependence. In: Wiley StatsRef: Statistics Reference Online (2014)
122. Scardapane, S., Wang, D.: Randomness in neural networks: An overview. In: Wiley Interdisciplinary Reviews: Data Mining and Knowledge Discovery, vol. 7(2) (2017)
123. Seshadri, V., West, B.J.: Fractal dimensionality of Lévy processes. Proc. Natl. Acad. Sci. U. S. A. **79**(14), 4501 (1982)
124. Shalalfeh, L., Bogdan, P., Jonckheere, E.: Fractional dynamics of PMU data. IEEE Trans. Smart Grid **12**(3), 2578–2588 (2020)

125. Sheng, H., Chen, Y.: FARIMA with stable innovations model of Great Salt Lake elevation time series. Signal Process. **91**(3), 553–561 (2011)
126. Sheng, H., Chen, Y., Qiu, T.: Fractional Processes and Fractional-order Signal Processing: Techniques and Applications. Springer Science & Business Media, New York (2011)
127. Sheng, H., Chen, Y.Q., Qiu, T.: Heavy-tailed distribution and local long memory in time series of molecular motion on the cell membrane. Fluctuation Noise Lett. **10**(01), 93–119 (2011)
128. Sheng, H., Sun, H., Chen, Y., Qiu, T.: Synthesis of multifractional Gaussian noises based on variable-order fractional operators. Signal Process. **91**(7), 1645–1650 (2011)
129. Spiegel, M.R., Schiller, J.J., Srinivasan, R.: Probability and Statistics. McGraw-Hill, New York (2013)
130. Sun, H., Chen, Y., Chen, W.: Random-order fractional differential equation models. Signal Process. **91**(3), 525–530 (2011)
131. Sun, R., Chen, Y., Zaveri, N., Zhou, A.: Local analysis of long range dependence based on fractional Fourier transform. In: Proceedings of the IEEE Mountain Workshop on Adaptive and Learning Systems, pp. 13–18. IEEE, New York (2006)
132. Sun, W., Li, Y., Li, C., Chen, Y.: Convergence speed of a fractional order consensus algorithm over undirected scale-free networks. Asian J. Control **13**(6), 936–946 (2011)
133. Swain, K.C., Thomson, S.J., Jayasuriya, H.P.: Adoption of an unmanned helicopter for low-altitude remote sensing to estimate yield and total biomass of a rice crop. Trans. ASABE **53**(1), 21–27 (2010)
134. Tanner, R.I., Walters, K.: Rheology: An Historical Perspective. Elsevier, Amsterdam (1998)
135. Tarasov, V.E.: Fractional vector calculus and fractional Maxwell's equations. Ann. Phys. **323**(11), 2756–2778 (2008)
136. Tarasov, V.E.: Fractional Dynamics: Applications of Fractional Calculus to Dynamics of Particles, Fields and Media. Springer Science & Business Media, New York (2011)
137. Tieleman, T., Hinton, G.: Divide the gradient by a running average of its recent magnitude. In: Coursera: Neural networks for machine learning (2017)
138. Tyukin, I.Y., Prokhorov, D.V.: Feasibility of random basis function approximators for modeling and control. In: Proceedings of the IEEE Control Applications, (CCA) & Intelligent Control, (ISIC) (2009)
139. Unser, M., Blu, T.: Fractional splines and wavelets. SIAM Rev. **42**(1), 43–67 (2000)
140. Valério, D., Machado, J., Kiryakova, V.: Some pioneers of the applications of fractional calculus. Fract. Calc. Appl. Anal. **17**(2), 552–578 (2014)
141. Vinagre, B.M., Chen, Y.: Lecture notes on fractional calculus applications in automatic control and robotics. In: Proceedings of the 41st IEEE CDC Tutorial Workshop, vol. 2, pp. 1–310 (2002)
142. Vinnicombe, G.: Uncertainty and Feedback: H_∞ Loop-shaping and the ν-gap Metric. World Scientific, Singapore (2001)
143. Viola, J., Chen, Y., Wang, J.: Information-based model discrimination for digital twin behavioral matching. In: Proceedings of the International Conference on Industrial Artificial Intelligence (IAI), pp. 1–6. IEEE, New York (2020)
144. Viswanathan, G.M., Afanasyev, V., Buldyrev, S., Murphy, E., Prince, P., Stanley, H.E.: Lévy flight search patterns of wandering albatrosses. Nature **381**(6581), 413–415 (1996)
145. Wang, D., Li, M.: Stochastic configuration networks: fundamentals and algorithms. IEEE Trans. Cybern. **47**(10), 3466–3479 (2017)
146. Wang, D., Li, M.: Deep stochastic configuration networks with universal approximation property. In: Proceedings of the International Joint Conference on Neural Networks (IJCNN), pp. 1–8. IEEE, New York (2018)
147. Wang, W., Wang, D.: Prediction of component concentrations in sodium aluminate liquor using stochastic configuration networks. Neural Comput. Appl. **32**(17), 13625–13638 (2020)
148. Ward, J.S., Barker, A.: Undefined by data: a survey of big data definitions. arXiv preprint arXiv:1309.5821 (2013)
149. Wei, J.: Research on swarm intelligence optimization algorithms and their applications to parameter identification of fractional-order systems. Ph.D. thesis, Beijing Jiaotong University, Beijing (2020)

150. Wei, J., Chen, Y., Yu, Y., Chen, Y.: Improving cuckoo search algorithm with Mittag-Leffler distribution. In: Proceedings of the International Design Engineering Technical Conferences and Computers and Information in Engineering Conference, pp. 1–9. American Society of Mechanical Engineers, New York (2019)
151. Wei, J., Chen, Y., Yu, Y., Chen, Y.: Optimal randomness in swarm-based search. Mathematics **7**(9), 828 (2019)
152. Wei, J., Yu, Y.: An adaptive cuckoo search algorithm with optional external archive for global numerical optimization. In: Proceedings of the International Conference on Fractional Differentiation and its Applications (ICFDA) (2018)
153. Wei, J., Yu, Y.: A novel cuckoo search algorithm under adaptive parameter control for global numerical optimization. Soft Comput. **24**, 4917–4940 (2019)
154. West, B.J.: Fractional Calculus View of Complexity: Tomorrow's Science. CRC Press, New York (2016)
155. West, B.J.: Sir Isaac Newton stranger in a strange land. Entropy **22**(11), 1204 (2020)
156. West, B.J., Geneston, E.L., Grigolini, P.: Maximizing information exchange between complex networks. Phys. Rep. **468**(1–3), 1–99 (2008)
157. West, B.J., Grigolini, P.: Complex Webs: Anticipating the Improbable. Cambridge University, Cambridge (2010)
158. Wilson, A.C., Recht, B., Jordan, M.I.: A Lyapunov analysis of momentum methods in optimization. arXiv preprint arXiv:1611.02635 (2016)
159. Wolpert, R.L., Taqqu, M.S.: Fractional Ornstein–Uhlenbeck Lévy processes and the telecom process: upstairs and downstairs. Signal Process. **85**(8), 1523–1545 (2005)
160. Woodward, W.A., Cheng, Q.C., Gray, H.L.: A *k*-factor GARMA long-memory model. J. Time Ser. Anal. **19**(4), 485–504 (1998)
161. Xue, D., Chen, Y.: Solving Applied Mathematical Problems with MATLAB. CRC Press, New York (2009)
162. Yang, Q., Chen, D., Zhao, T., Chen, Y.: Fractional calculus in image processing: a review. Fract. Calc. Appl. Anal. **19**(5), 1222–1249 (2016)
163. Yang, X.S.: Nature-inspired Metaheuristic Algorithms. Luniver Press, United Kingdom (2010)
164. Yang, X.S., Deb, S.: Engineering optimisation by cuckoo search. Int. J. Math. Model. Numer. Optim. **1**(4), 330–343 (2010)
165. Zadeh, L.A.: Fuzzy logic. Computer **21**(4), 83–93 (1988)
166. Zarco-Tejada, P.J., González-Dugo, V., Williams, L., Suárez, L., Berni, J.A., Goldhamer, D., Fereres, E.: A PRI-based water stress index combining structural and chlorophyll effects: assessment using diurnal narrow-band airborne imagery and the CWSI thermal index. Remote Sens. Environ. **138**, 38–50 (2013)
167. Zaslavsky, G.M., Sagdeev, R., Usikov, D., Chernikov, A.: Weak Chaos and Quasi-regular Patterns. Cambridge University Press, Cambridge (1992)
168. Zeng, C., Chen, Y.: Optimal random search, fractional dynamics and fractional calculus. Fract. Calc. Appl. Anal. **17**(2), 321–332 (2014)
169. Zhao, J., Yu, H., Luo, J.H., Cao, Z.W., Li, Y.X.: Hierarchical modularity of nested bow-ties in metabolic networks. BMC Bioinf. **7**(1), 1–16 (2006)
170. Zhao, T., Chen, Y., Ray, A., Doll, D.: Quantifying almond water stress using unmanned aerial vehicles (UAVs): correlation of stem water potential and higher order moments of non-normalized canopy distribution. In: ASME 2017 International Design Engineering Technical Conferences and Computers and Information in Engineering Conference. American Society of Mechanical Engineers, New York (2017)
171. Zhao, T., Niu, H., de la Rosa, E., Doll, D., Wang, D., Chen, Y.: Tree canopy differentiation using instance-aware semantic segmentation. In: 2018 ASABE Annual International Meeting. American Society of Agricultural and Biological Engineers (2018)
172. Zhao, Z., Guo, Q., Li, C.: A fractional model for the allometric scaling laws. Open Appl. Math. J. **2**(1), 26–30 (2008)

Part II
Smart Big Data Acquisition Platforms

Chapter 3
Small Unmanned Aerial Vehicles (UAVs) and Remote Sensing Payloads

Abstract This chapter navigates the realm of small Unmanned Aerial Vehicles (UAVs) and their integral role in remote sensing applications, focusing on the sophisticated payloads they carry. The exploration begins with an in-depth examination of the UAV platform, elucidating the key features that make it an invaluable tool for aerial data collection. Within the context of lightweight sensors, the chapter outlines the capabilities of various sensors, including RGB cameras, multispectral cameras, short-wave infrared cameras, and thermal cameras. The discussion extends to UAV image acquisition and processing, covering flight mission design and the intricacies of image processing techniques. Challenges and opportunities in the utilization of UAVs and associated image processing are meticulously explored, encompassing aspects such as UAV path planning, pre-flight path planning, multispectral image calibration, thermal camera calibration, and image stitching for orthomosaic generation. A compelling case study is presented, investigating the impact of high spatial resolution on the Normalized Difference Vegetation Index (NDVI) in UAV-based individual tree-level mapping. The study employs evidence from nine field tests and discusses implications, providing valuable insights into the practical implications of UAV remote sensing at a high spatial resolution.

3.1 The UAV Platform

Many kinds of UAVs are used for different research purposes, such as evapotranspiration estimation. Some popular UAV platforms are shown in Fig. 3.1. Typically, there are two types of UAV platforms, fixed-wings and multirotors. Fixed-wings can usually fly longer with a larger payload. It can usually fly for about 2 hours, which is suitable for a large field. Multirotors can fly about 30 minutes with payload, which is suitable for short flight missions. Both of them have been used in agricultural research, such as [51, 54], which promises great potential in digital agriculture.

The authors mainly used a quadcopter named "Hover" to collect aerial images, as shown in Fig. 3.1e. The "Hover" was equipped with a Pixhawk flight controller, GPS, telemetry antennas. It can fly over the field by waypoints mode (designed by using Mission Planner software). The lithium polymer battery of "Hover" has a

H. Niu, Y. Chen, *Smart Big Data in Digital Agriculture Applications*, Agriculture Automation and Control, https://doi.org/10.1007/978-3-031-52645-9_3

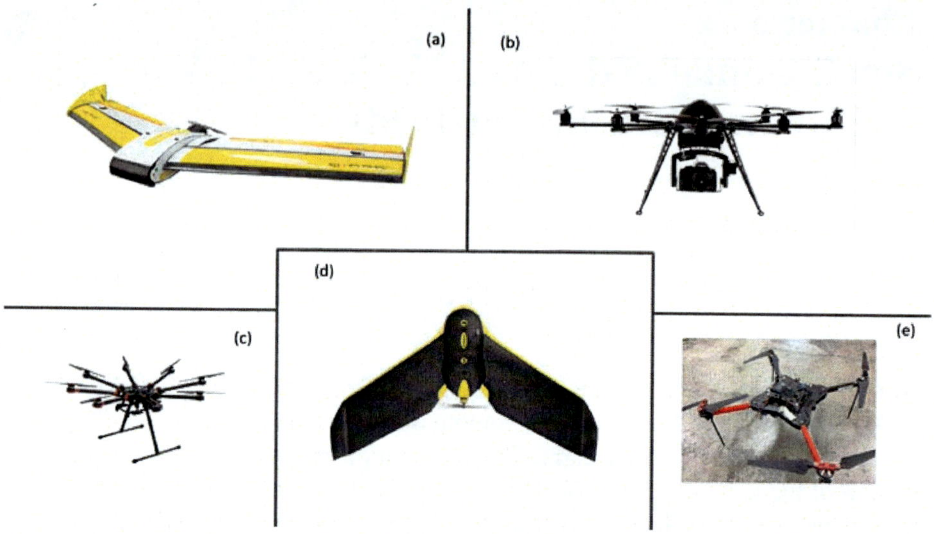

Fig. 3.1 (**a**) The QuestUAV 200 UAV. (**b**) The MK Okto XL 6S12. (**c**) The DJI S1000. (**d**) The eBee Classic. (**e**) The Hover

Table 3.1 The specifications of "Hover." The quadcopter is equipped with high efficient power system, including T-Motor MN3508 KV380 motor, 1552 folding propeller and Foxtech Multi-Pal 40A OPTP ESC, to ensure long flight time

Specifications	
Wheelbase	610 mm
Folding size	285 × 285 × 175 mm
Propeller	Foxtech 1552 folding propeller
Motor	T-Motor MN3508 KV380
ESC	Foxtech Multi-Pal 40A OPTO ESC (Simonk Firmware)
Flight controller	Pixhawk cube orange standard set with Here 2 GNSS
Operating temperature	$-20\sim+50\,^{\circ}\mathrm{C}$
Suggested flight altitude	<1000 m
Max air speed	20 m/s

capacity of 9500 mAh, which can support a 30-minute flight mission with cameras mounted on it. The specifications of the "Hover" are listed in Table 3.1 for reference. The "Hover" is equipped with high efficient power system, including T-Motor MN3508 KV380 motor, 1552 folding propeller, and Foxtech Multi-Pal 40A OPTP ESC, to ensure long flight time.

Fig. 3.2 The Survey 2 sensors and the reflectance calibration ground target package

3.2 Lightweight Sensors

Mounted on UAVs, many light weight sensors can be used for collecting UAV imagery, such as RGB, multispectral, and thermal images, for agricultural research. In this section, the authors listed the sensors that had been commonly used in most of the research work reported in this monograph. The sensors being introduced here will be mentioned in the following chapters frequently. Therefore, the authors introduced the sensors in this section in detail.

3.2.1 RGB Camera

The Survey 2 (MAPIR, San Diego, CA, USA)[1] camera has four bands, Blue, Green, Red, and Near-infrared (NIR), with a spectral resolution of 4608 × 3456 pixels, and a spatial resolution of 1.01 cm/pixel (Fig. 3.2). The Survey 2 camera has a fast interval timer, 2 seconds for JPG mode, and 3 seconds for RAW + JPG mode. Faster interval timer would benefit the overlap percentage design for UAV flight missions, such as reducing the flight time and enabling higher overlapping ratio.

[1] Mention of trade names or commercial products in this publication is solely for the purpose of providing specific information and does not imply recommendation or endorsement by the University of California. The University of California is equal opportunity providers and employers.

Fig. 3.3 The RedEdge M is a commonly used multispectral camera. The RedEdge M has five bands, which are Blue, Green, Red, Near-infrared, and Red edge. It has a spectral resolution of 1280 × 960 pixels, with a 46° field of view

3.2.2 *Multispectral Camera*

The RedEdge M is a commonly used multispectral camera (Fig. 3.3). The RedEdge M has five bands, which are Blue, Green, Red, Near-infrared, and Red edge. It has a spectral resolution of 1280 × 960 pixels, with a 46° field of view. With a Downwelling Light Sensor (DLS), which is a five-band light sensor that connects to the camera, the RedEdge M can measure the ambient light during a flight mission for each of the five bands. Then, it can record the light information in the metadata of the images captured by the camera. After the camera calibration, the information detected by the DLS can be used to correct lighting changes during a flight, such as changes in cloud cover during a UAV flight.

3.2.3 *The Short Wave Infrared Camera*

The SWIR 640 P-Series (Infrared Cameras Inc, Beaumont, TX, USA), which is a short-wave infrared camera, has also been commonly used for agricultural research (Fig. 3.4). The spectral band is from 0.9 μm to 1.7 μm. The accuracy for the SWIR camera is ±1 °C. It has a resolution of 640 × 512 pixels.

3.2.4 *Thermal Camera*

The thermal camera ICI 9640 P (Infrared Cameras Inc, Beaumont, TX, USA) has been used for collecting thermal images as reported in [29, 35, 40, 50]. The thermal camera has a resolution of 640 × 480 pixels. The spectral band is from 7 μm to 14 μm. The dimension of the thermal camera is 34 mm × 30 mm × 34 mm (Fig. 3.5). The accuracy is designed to be ±2 °C. A Raspberry Pi Model B computer (Raspberry Pi Foundation, Cambridge, UK) can be used to trigger the thermal camera during flight missions.

Fig. 3.4 The SWIR 640 P-Series, which is a short-wave infrared camera, has also been commonly used for agricultural research. The spectral band is from 0.9 μm to 1.7 μm. The accuracy for the SWIR camera is ± 1 °C. It has a resolution of 640 $\times$ 512 pixels

3.3 UAV Image Acquisition and Processing

3.3.1 Flight Mission Design

The authors used the Mission Planner to program all flight missions (Fig. 3.6). The flight height was usually set up as 30, 60, 90, and 120 m based on research purpose. The overlapping of UAV imagery was set up as 80%, so that the UAV imagery of the crops or trees can be stitched together during image processing. A bi-weekly UAV flight schedule is suggested to collect sufficient data. If there is a UAV crash, unexpected weather conditions, hardware issues, or unknown reasons, data may not be collected successfully. If data is missed, people may have to wait for another year.

To minimize the shading effect on the images, the UAVs are usually flying at noon with clear sky conditions. Because each pixel in a UAV image is a percentage of the reflected light, pixel values need to be calibrated by using a known reflectance value. Therefore, the image of a calibration board needs to be taken before and after the flight missions, servicing as the reflectance reference. It is important to take pictures of the reference panel immediately before and after the flight missions because the solar angle and light intensity can change [52], which causes inaccurate experiment results. UAV images usually have higher radiometric homogeneity than aircraft or satellite images because of the lower flight altitude [18]. However, there are also special UAVs image quality problems. For example, the camera position on the UAVs might be different for each flight mission, which can cause different

Fig. 3.5 The thermal camera has a resolution of 640 × 480 pixels. The spectral band is from 7 μm to 14 μm. The dimension of the thermal camera is 34 mm × 30 mm × 34 mm. The accuracy is designed to be ±2 °C

spatial resolution or different viewing angles [18]. The low flight height of UAVs can also result in geometric distortion [18, 45]. Besides, lower flight height results in greater numbers of UAV images to keep effective overlapping, which makes image processing more time-consuming.

3.3.2 UAV Image Processing

After the flight missions, all of the aerial images were stitched together to generate the orthomosaic images (Table 3.2 and Fig. 3.7) in Metashape (Agisoft LLC,

Fig. 3.6 The user interface of Mission Planner. The example of flight mission was for nematode data collection using UAV for Project 30 at UC Kearny Center. See Chap. 7 for more details

Russian). Preselection is recommended because it can speed up the processing of large datasets. Building the dense cloud can reconstruct a more accurate surface, which can improve the quality of the final orthomosaic. Higher quality usually can result in a more accurate surface, which means a greater number of points. However, higher quality is not recommended because of longer data processing time. Medium quality is sufficient for UAVs image processing, especially for low variations field. Building Digital Elevation Model (DEM) allows generating an accurate surface, which can be used as a source for the orthomosaic generation. This will shorten the data processing time compared with Build Mesh operation because Build Mesh is usually used for a more complex surface. The source data for building DEM is the dense cloud. For the interpolation method, **Extrapolated** option is selected because it can generate a surface without gaps being extrapolated to the bound box sides. The default option for **Interpolation** is **Enabled**, which is not recommended because it will leave the valid elevation values only for fields that are seen from at least one aligned camera.

3.4 Challenges and Opportunities

Compared with traditional remote sensing tools, such as satellite, the UAVs flight can be more flexible and frequent in the field. UAVs can fly at a lower altitude and can take higher spatial and temporal resolution images of crops [50]. As a low-cost scientific data collection platform, UAVs also make data acquisition relatively less expensive. While there are many advantages by using UAVs for agricultural

Table 3.2 Orthomosaic images generation workflow in Agisoft Metashape

Step 1: Align Photos	Step 2: Build Mesh	Step 3: Build Orthomosaick
Accuracy: Medium	Surface type: Height field (2.5D)	Type: Planar
Generic preselection: Yes	Source data: Sparse cloud	Projection plane: TOP XY
Key point limit: 40,000	Face count: Medium (30,000)	Rotation angle: 0
Tie point limit: 4000	Interpolation: Enabled (default)	Surface: Mesh
Adaptive camera model fitting: No	Point classes: All	Blending mode: Mosaic (default)
	Calculate vertex colors: Yes	Enable hole filling: Yes
		Enable back-face culling: No

Fig. 3.7 Agisoft Metashape image processing workflow: (**a**) Align Photos. (**b**) Build Mesh. (**c**) Generate orthomosaick

research, there are still challenges for UAVs. These challenges are also commonly shown in different research work [3, 5, 6, 13, 21, 22, 24, 25, 30, 31, 33, 44].

3.4.1 UAVs

Although UAVs have shown great potential for digital agriculture, there are still many technical problems for UAVs, such as flight time, flight height control, path planning, stability in winds, and turbulence [10, 17]. For example, most UAVs can only fly around 30 minutes with payload, which is not enough for a large field. The power can also run low faster because of unexpected headwinds or other factors. Increasing the payload of UAVs will make the UAVs more capable. Flight height is another concern, in the United States, the maximum altitude for UAVs is limited to 120 m. The UAVs need to be in the sight of the operator, and a pilot license is also required. Consequently, it is necessary to have a flying team for UAVs. For a detailed discussion on technical limitations for UAVs, please refer to [11]. Fortunately, it is expected that with the development of UAV technology, new camera designs, lower costs, improved image processing techniques, and a greater number of experimental studies of UAV-based remote sensing for agriculture applications, UAVs will have better performance for agricultural research.

3.4.2 UAV Path Planning and Image Processing

Many researchers fly the UAVs in different height, using specialized equipment, controlling environments, and relying on data analysis expertise [38]. Is there any optimal set of parameters such that the data can be the best representation of crops or trees? In [38], Stark et al. built a conceptual framework for describing the optimality as a function of spatial, spectral, and temporal factors that represent the best solution. As researchers try to understand the potential of the UAVs, efficient workflow, image processing methods, and better software are still under developing [12, 19, 42, 43].

3.4.3 Pre-flight Path Planning

Being used as a remote sensing platform, UAVs also create new research problems, such as UAVs image processing and flight path planning. Flight missions are usually designed by different kinds of software. The flight height is usually set up as 30 m, 60 m, and 120 m. For all the flight missions in the reviewed papers, the overlap was usually set up between 75% and 85% to enable the images stitched together during image processing. The UAVs sensors are designed to take images at nearly 0 nadir angle.

Researchers usually fly UAVs biweekly to collect data. If there is a UAV crash, unexpected weather conditions, hardware issues, or unknown reasons, data may not be collected successfully. If data is missed, people may have to wait for another year. A bi-weekly UAV flight schedule is suggested to collect sufficient data.

3.4.4 Multispectral Image Calibration

To minimize the shading effect on the images, the UAVs are usually flying at noon with clear sky conditions. Because each pixel in a UAV image is a percentage of the reflected light, pixel values need to be calibrated by using a known reflectance

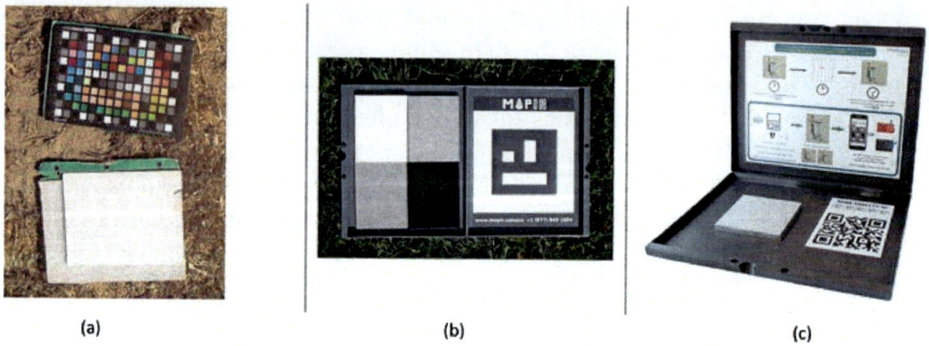

Fig. 3.8 The UAV image reflectance calibration. (**a**) A color panel. (**b**) The MAPIR target calibration board. (**c**) The RedEdge M calibration board

value. Therefore, the image of a calibration board needs to be taken before and after the flight missions, serving as the reflectance reference (Fig. 3.8).

It is important to take pictures of the reference panel immediately before and after the flight missions because the solar angle and light intensity can change [52], which causes inaccurate experiment results. The digital number of the images are converted to reflectance by Smith and Milton [37]

$$\rho_\lambda = \frac{DN - DN_d}{DN_w - DN_d}, \tag{3.1}$$

where ρ_λ is the reflectance and DN is the digital number of the raw image, DN_d and DN_w are the dark reflectance point and white reflectance point in the color checker, respectively.

UAV images usually have higher radiometric homogeneity than aircraft or satellite images because of the lower flight altitude [18]. However, there are also special UAV image quality problems. For example, the camera position on the UAVs might be different for each flight mission, which can cause different spatial resolution or different viewing angles [18]. The low flight height of UAVs can also result in geometric distortion [18, 45]. Besides, lower flight height results in greater numbers of UAV images to keep effective overlapping, which makes image processing more time-consuming.

Although multispectral cameras have light sensors to calibrate light conditions, saturation issues can still be found in UAV images. For example, with a Downwelling Light Sensor (DLS), which is a five-band sensor that connects to the multispectral camera, the RedEdge M can measure the ambient light during a flight and record the light information in the images. After the camera calibration, the information detected by the DLS can be used to correct lighting changes during a flight, which usually happens because the clouds cover the sun during a UAV flight. The clouds are believed to affect the saturation issues, even though sunshine is supposed to correct reflectance for real-time conditions. Saturated values decrease the quality of the data. The retrieval of required indexes, such as NDVI and LAI, are important for the estimation of soil heat flux G and sensible heat flux H.

Another critical issue with UAVs is the bidirectional reflectance distribution function (BRDF) effects. For many UAV application, the reflectance model for canopy measurements is simplified to assume a strict nadir (or straight down) viewing angle and a static illumination source [2, 39, 52]. However, this assumption does not consider the bidirectional reflectance distribution function (BRDF). The BRDF is a function of wavelength, observer azimuth, observer zenith, illumination azimuth, and illumination zenith [39]. In satellite images, the effect of BRDF is relatively uniform because the satellite covers a wide region in a single frame. However, this simplification is not valid for UAV platforms equipped with an imaging system with a wide field of view (FOV).

Further experiment should be based on multispectral measurements, and UAV image acquisition should be conducted to select those spectral bands which are most useful and sensitive for specific research purpose. Cameras should be designed only for those needed bands, which will lower the cost of the sensors. The availability of low-cost UAV platforms and specialized cameras will also make the UAV application on agriculture more competitive.

3.4.5 Thermal Camera Calibration and Image Processing

The thermal image from UAVs is becoming a useful source for agricultural research because of the higher temporal and spatial resolution compared with those obtained from the satellite. The thermal camera has a spectral response from 7 μm to 14 μm. The accuracy can be as high as $\pm 1\,^{\circ}$C. Thermal remote sensing images were first used in 1973 to estimate ET [4]. Temperature information is usually converted into land surface characteristics such as albedo, LAI, and surface emissivity. The TIR band is considered as the most critical variable for estimating the sensible heat flux and ground heat flux [6]. The cooled thermal cameras are usually more sensitive and accurate than uncooled thermal cameras [36]. However, cooled thermal cameras are very big, expensive, and energy-consuming [35]. Therefore, cooled thermal cameras can hardly be used on UAV platforms. In contrast, the uncooled thermal cameras are usually lighter [2], which are usually less than 200 grams, low power consumption [8], and less expensive than cooled thermal cameras.

The thermal camera has many advantages, though, its microbolometer is not always sensitive and accurate [35]. Most thermal cameras are not always calibrated, which can only measure the relative temperature instead of the absolute values. In digital agriculture, it is necessary to measure the absolute temperature in many applications [2], such as crop monitoring [15], pest detection [9], and disease detection [20]. Unstable outdoor environmental factors can also cause serious measurement drift during flight missions. Post-processing like mosaicking might further lead to measurement errors. To solve these two fundamental problems, in [50], the authors conducted three experiments to research the best practice of thermal image collection using UAVs. To calibrate TIR images, in [31], Park et al. used the water body and rubber plates as cold and hot features. IR Flash Version 2 is usually used to process thermal UAV images for image format transformation.

The correlation between the measured IR temperature from calibration boards and the estimates by thermal cameras were found to be unacceptable sometimes. Without warming up the uncooled thermal camera, the temperature difference between the thermal camera and calibration board can be as high as ±10 °C. For instance, the land surface temperature is the most important data for SEBAL and the estimation of surface energy fluxes; thus, its accuracy is the key for a reliable ET estimation.

Many researchers also focus on thermal camera calibration issues. For example, Ribeiro et al. proposed a new calibration algorithm based on neural networks [35]. The calibration algorithms considered the thermal camera temperature and the digital response of the microbolometer as input data. Based on the calibration result, the accuracy increased from 3.55 °C to 1.37 °C. In [40], Torres-Rua et al. presented a vicarious calibration methodology (UAV-specific, time-specific, flight-specific, and sensor-specific) for thermal camera images traceable back to NIST-standards (National Institute of Standards and Technology) and current atmospheric correction methods.

For future research, uncooled thermal cameras can be used to evaluate with other temperatures sensor information to acquire reliable thermal information, such as atmospherically corrected satellite images and temperature canopy sensors.

3.4.6 Images Stitching and Orthomosaic Image Generation

After UAV images are collected, all of the aerial images need to be stitched together to generate the orthomosaic images. Some problems are identified when creating mosaics, such as fault lines, blurriness, and replicated features, especially with the thermal data. To overcome the thermal camera's effect, a higher overlap for the thermal camera can be a good choice. With an increase in the image overlap by 5%, most of the fuzziness and replicated problems were eliminated [21].

There are many types of software that can be used for image stitching, such as Pix4D (Pix4D, Prilly, Switzerland), Agisoft Metashape, RealityCapture, and DroneDeploy (DroneDeploy, San Francisco, CA, USA). For example, during the image stitching workflow using the Agisoft Metashape, there are several steps for image processing, which include aligning photos, optimizing cameras, building mesh, building dense cloud, building digital elevation model (DEM), and generating orthomosaic.

3.5 Case Study: High Spatial Resolution Has Little Impact on NDVI Mean Value of UAV-Based Individual Tree-level Mapping: Evidence from Nine Field Tests and Implications

3.5.1 Introduction

The normalized difference vegetation index (NDVI) has been used for many agriculture-related research topics, such as water stress detection [48, 52], crop yield assessment [53], and ET estimation [26, 28]. The NDVI is usually calculated by

$$NDVI = \frac{\rho_{nir} - \rho_r}{\rho_{nir} + \rho_r}, \tag{3.2}$$

where ρ_{nir} and ρ_r are the reflectances of the near-infrared and red wavebands, respectively. NDVI is a standardized method to measure healthy vegetation. When the NDVI is high, it indicates the vegetation has a higher level of photosynthesis.

To date, satellite-derived NDVI has been commonly used for crop coefficient values estimation [14, 16, 34]. For example, Trout et al. [41] and Zhang et al. [46] applied a remote sensing method using NDVI to estimate canopy ground cover as a basis for generating crop coefficient K_c. Kamble et al. [16] used a simple linear regression model to establish a relationship between NDVI and K_c. Although satellite imagery can obtain spatially distributed measurements, they cannot acquire high spatio-temporal resolution images for digital agriculture applications [7]. The timing of satellite overpass is not always synchronous with research requirements, either.

With the development of new remote sensing technology, the unmanned aerial vehicles (UAVs) have been commonly used in agricultural applications, such as crop yield estimation [53], irrigation managements [29, 49], water stress estimation [47], and pest management [23, 27]. Compared with the satellite, the flight of UAVs can be more flexible and frequent in the field. The UAVs fly at a lower altitude and take higher resolution imagery of crops [50]. The UAVs also make data acquisition relatively less expensive. However, one may fly the UAV at different flight heights. What is the optimal UAV flight height for research can be an interesting topic. In previous paper [1], Awais et al. investigated the optimal timing and altitude for thermal imagery collection using UAV in an Anji tea plant experimental field. The results reported that the thermal imagery could provide the best correlation and accurate canopy temperature when the UAV flights were at 11 am and 60 m altitude. In previous article [29], the authors applied supervised learning methods to study the correlation between imagery features and onions irrigation treatments. Then, it figured out how UAV flight height or resolution settings affect the accuracy of onion irrigation treatment inference.

Studies of [1, 29] showed the importance of UAV flight height or spatial resolution's effect on data analysis. To date, few studies have investigated the

Fig. 3.9 The pomegranate study site at USDA. The pomegranate was planted in 2010 with a 5 m spacing between rows and 2.75 m within-row tree spacing in a 1.3 ha field. There are two large weighing lysimeters located in the center of the field, marked in red boxes

association between NDVI and UAV flight height or spatial resolution at **individual tree level**. The **objective of this study** are: (1) Investigate how the UAV flight height or spatial resolution affects the mean NDVI for individual tree-level canopy. (2) Check the reliability of the multispectral sensor for different heights of UAV flight missions. The **major contribution of this section** are: (1) Publish a dataset on Dryad for a high-spatial resolution UAV imagery research study. All the datasets will be available at DRYAD[2] for research purpose. (2) Point out the importance of variability analysis of individual tree-level research.

The rest of this section was organized as follows. Section 3.2 introduced the material and methods used to calculate the mean NDVI of the individual tree canopy. Results and discussion were presented in Sect. 3.3. In Sect. 3.4, concluding remarks were presented.

3.5.2 Material and Methods

3.5.2.1 The Study Site

The flight missions were conducted in a pomegranate field (Fig. 3.9) at the USDA-ARS, San Joaquin Valley Agricultural Sciences Center (36.594 °N, 119.512 °W), Parlier, California, 93648, USA.

[2] All the datasets will be available at https://doi.org/10.6071/M3JH4Q for research purpose. Citation: Niu, Haoyu; Chen, YangQuan (2021), RedEdge M pomegranate field 60 m, 90 m, 120 m, Dryad, Dataset, https://doi.org/10.6071/M3JH4Q.

3.5.2.2 The UAV and the Multispectral Sensor

In this article, the unmanned aerial vehicle (UAV) platform, named Hover, was adopted to conduct this exploratory study. The authors chose the RedEdge M camera (MicaSense, Seattle, WA, USA) to obtain multispectral imagery. The multispectral sensor has five different bands, which are Blue (475 nm), Green (560 nm), Red (668 nm), Near-infrared (840 nm), and Red edge (717 nm). The RedEdge M has a spectral resolution of 8.2 cm/pixel (per band) at 120 m (400 ft.) above ground level (AGL), with a 46° field of view.

3.5.2.3 Details of the UAV Imagery Dataset

The UAV flight height was 60 m, 90 m, and 120 m to acquire different high-resolution multispectral imagery. Data was collected successfully (Table 3.3) for three different days, 5-8-19, 9-19-19, and 10-3-19. All of the multispectral images were then processed to generate the orthomosaick images in Metashape (Agisoft LLC, Russian).

For image processing in Metashape, preselection was recommended because it could speed up large datasets processing. Building the dense cloud could reconstruct a more accurate surface, improving the quality of the final orthomosaic. Higher quality usually could result in a more accurate surface, which meant a greater number of points. However, higher quality was not recommended because of the longer data processing time. Medium quality was sufficient for UAVs image processing, especially for low variations field. Building Digital Elevation Model (DEM) allowed generating an accurate surface, which could be used as a source for the orthomosaic generation. This would shorten the data processing time compared with Build Mesh operation because Build Mesh was usually used for a more complex surface.

The source data for building DEM was the dense cloud. For the interpolation method, **Extrapolated** option was selected because it could generate a surface without gaps being extrapolated to the bound box sides. The default option for **Interpolation** was **Enabled**, which was not recommended because it would leave the valid elevation values only for fields seen from at least one aligned camera.

Table 3.3 The UAV flight schedule. The UAV flight height was 60 m, 90 m, and 120 m to acquire different high-resolution multispectral imagery. Data was collected successfully for three different days, 5-8-19, 9-19-19, and 10-3-19

Dates	Flight time	Flight height
May 8th, 2019	12–1 p.m.	60 m, 90 m and 120 m
Sep 19th, 2019	12–1 p.m.	60 m, 90 m and 120 m
Oct 3rd, 2019	12–1 p.m.	60 m, 90 m and 120 m

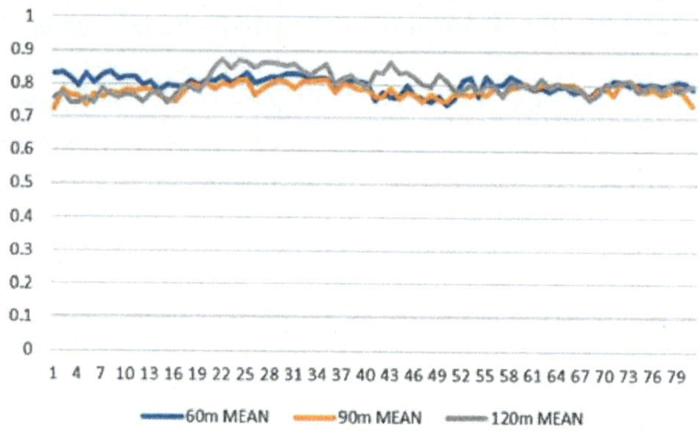

Fig. 3.10 The mean NDVI values of each sampling tree at 60 m, 90 m, and 120 m on May 8th, 2019. The *x*-axis is the identification number (ID) for sampling trees, 82 in total. The *y*-axis is the mean NDVI value for each tree canopy

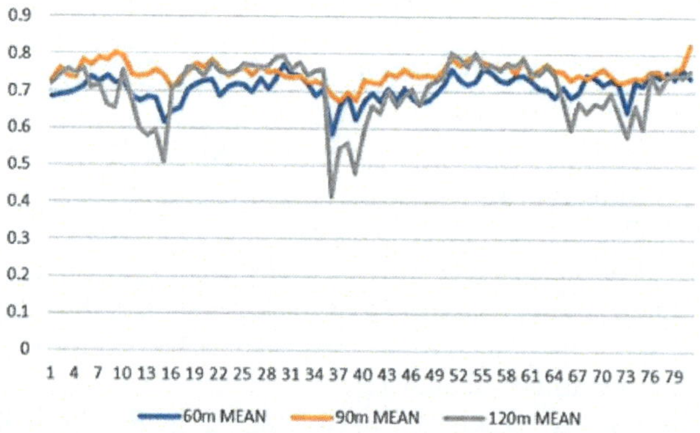

Fig. 3.11 The mean NDVI values of each sampling tree at 60 m, 90 m, and 120 m on Sep 19th, 2019. The *x*-axis is the identification number (ID) for sampling trees, 82 in total. The *y*-axis is the mean NDVI value for each tree canopy

3.5.3 *Results and Discussion*

3.5.3.1 The Relationship Between NDVI and UAV Flight Height

The mean NDVI values of each sampling tree were shown in Figs. 3.10, 3.11, and 3.12. Theoretically, for each sampling tree, the mean NDVI value of the tree canopy should have the same value at 60 m, 90 m, and 120 m. However, the values of NDVI could be very different from each other considering the weather conditions (Fig. 3.11), such as the cloud.

Key Observation In Fig. 3.11, the NDVI values were significantly different for trees from 1 to 20, and from 31 to 50. The reason was that for image segmentation, more shades were included in the tree canopy. In Figs. 3.10 and 3.12, the data was more consistent for different UAV flight height. For example, in Fig. 3.13, the authors compared the correlation of NDVI values between 90 m and 120 m. The result showed there was a strong correlation between them, with $R^2 = 0.7$.

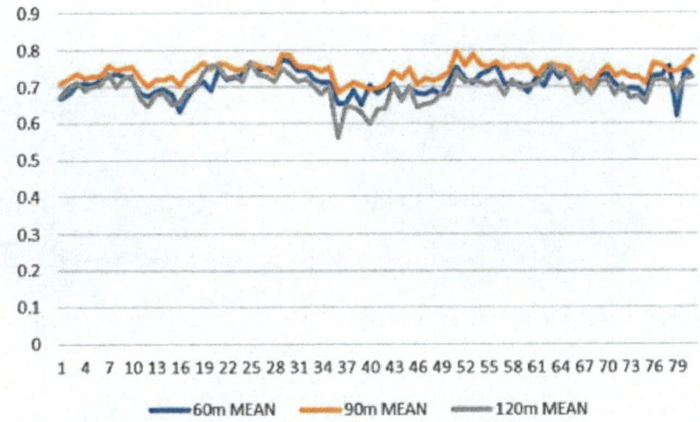

Fig. 3.12 The mean NDVI values of each sampling tree at 60 m, 90 m, and 120 m on Oct 3rd, 2019. The x-axis is the identification number (ID) for sampling trees, 82 in total. The y-axis is the mean NDVI value for each tree canopy

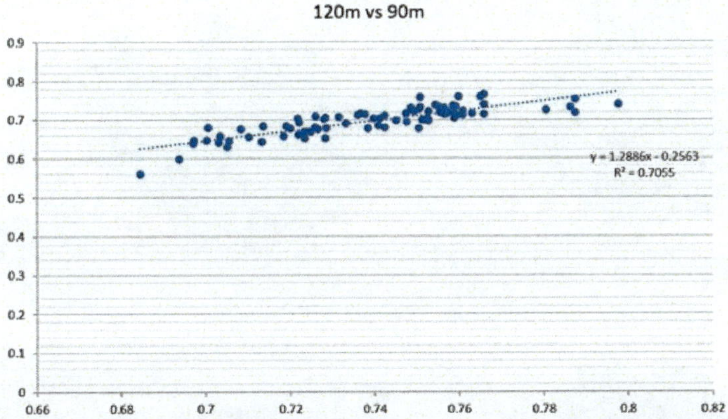

Fig. 3.13 The individual tree-level mean NDVI correlation between 120 m and 90 m on Oct 3rd, 2019. The x-axis is the mean NDVI values for sampling trees at 120 m flight height. The y-axis is the mean NDVI values for sampling trees at 90 m flight height

3.5.3.2 Individual Tree Canopy Segmentation Using Support Vector Machine (SVM)

To obtain the individual tree-level NDVI values of the 82 sampling trees, the authors used the SVM for classifying the tree canopy. Using the SVM classifier could map the input data vectors into a higher dimensional feature space. Then, the SVM optimally separated the data into different classes. Since the multispectral UAV imagery was large, the SVM classifier was adopted, which was less susceptible to noise, correlated bands, and unbalanced number or size of training sites within each class. All the sampling trees were successfully segmented using the SVM classifier.

Key Observation For simplicity, the authors only created the NDVI distribution for the two trees in lysimeter (Figs. 3.14, 3.15, and 3.16). For example, in Fig. 3.14, the NDVI distributions for the two trees in the lysimeter were generated. The NDVI was for May 8th, and the UAV flight height was at 60 m, 90 m, and 120 m. The color bar meant the range of the NDVI values (from −1 to 1). Based on the above section,

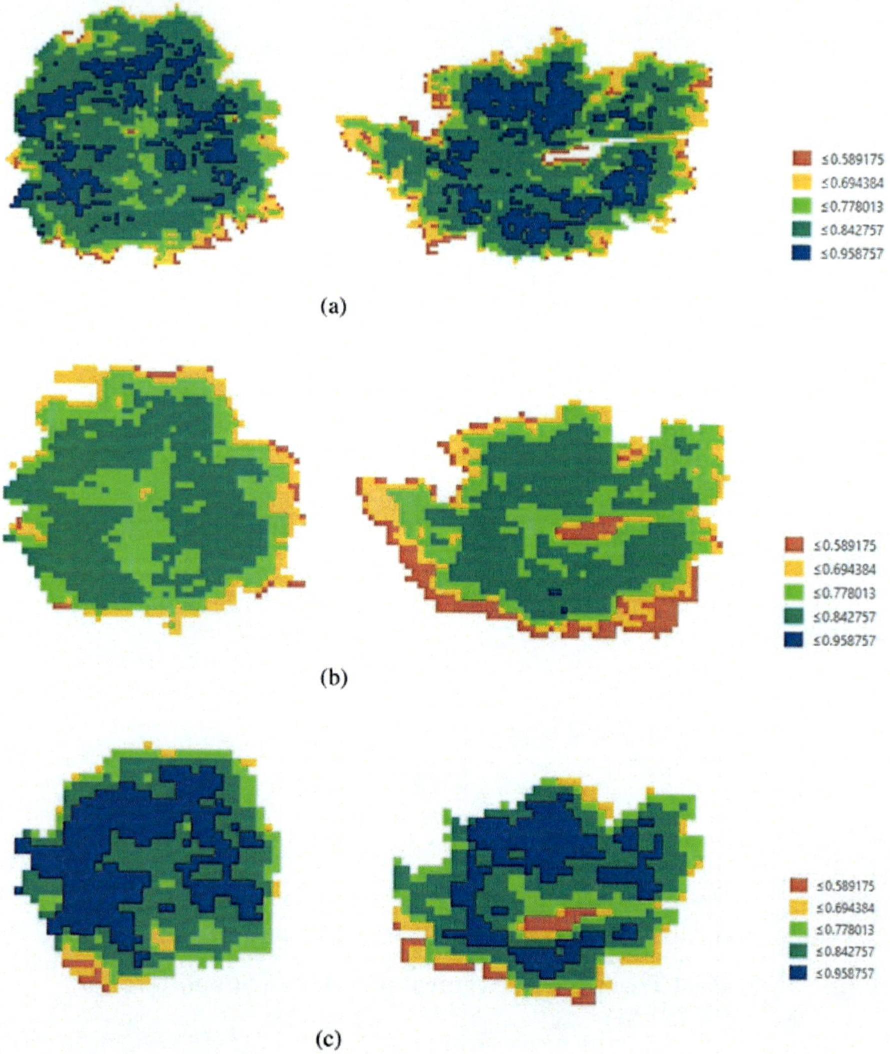

Fig. 3.14 The NDVI distribution of two individual lysimeter trees at 60 m, 90 m, and 120 m on May 8th, 2019. (**a**) The NDVI distribution of individual lysimeter trees at 60 m flight height. (**b**) The NDVI distribution of individual lysimeter trees at 90 m flight height. (**c**) The NDVI distribution of individual lysimeter trees at 120 m flight height

"The relationship between NDVI and UAV flight height," there was no significant difference for mean NDVI value at individual tree level. However, as shown in Fig. 3.14, lower flight height (60 m) gave a higher spatial resolution image. The NDVI distribution inside the canopy was more precise than that in higher flight height. Therefore, what the average told us could be wrong. How to use this high-resolution benefit from UAV imagery can be discussed in the near future. Variability analysis may play an important role in individual tree-level research.

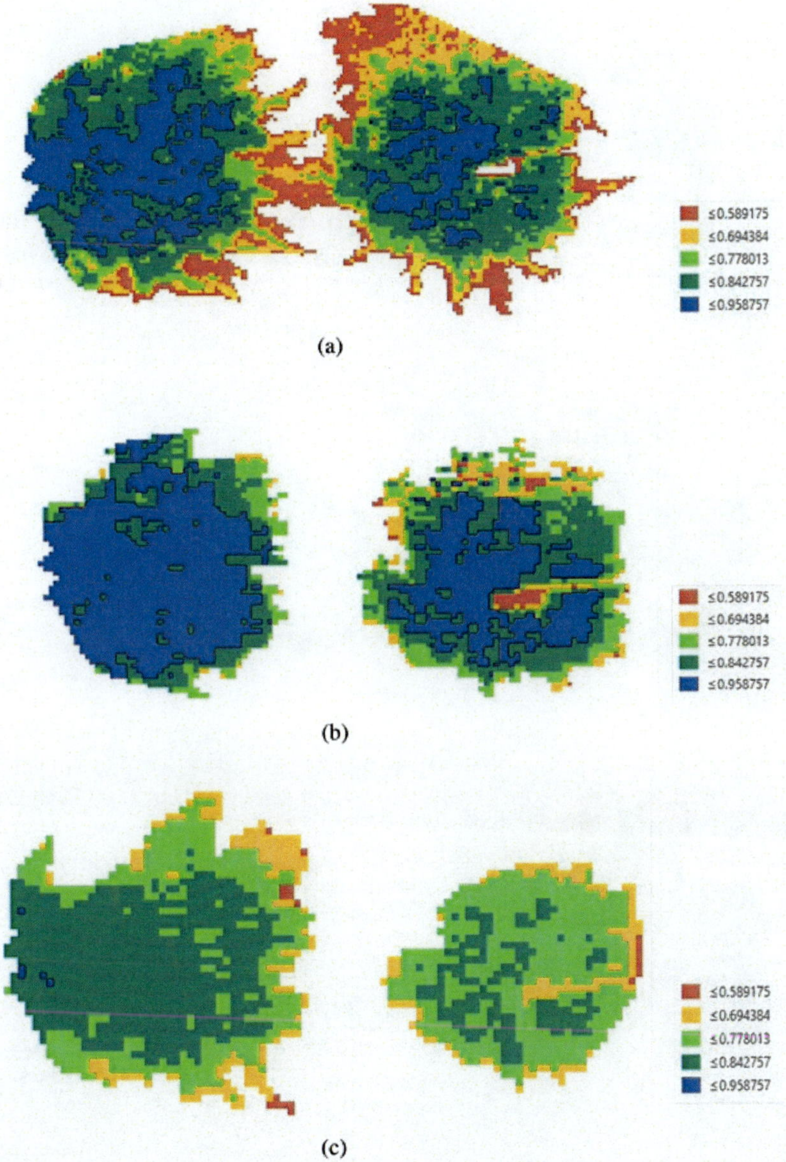

Fig. 3.15 The NDVI distribution of two individual lysimeter trees at 60 m, 90 m, and 120 m on Sep 19th, 2019. (**a**) The NDVI distribution of individual lysimeter trees at 60 m flight height. (**b**) The NDVI distribution of individual lysimeter trees at 90 m flight height. (**c**) The NDVI distribution of individual lysimeter trees at 120 m flight height

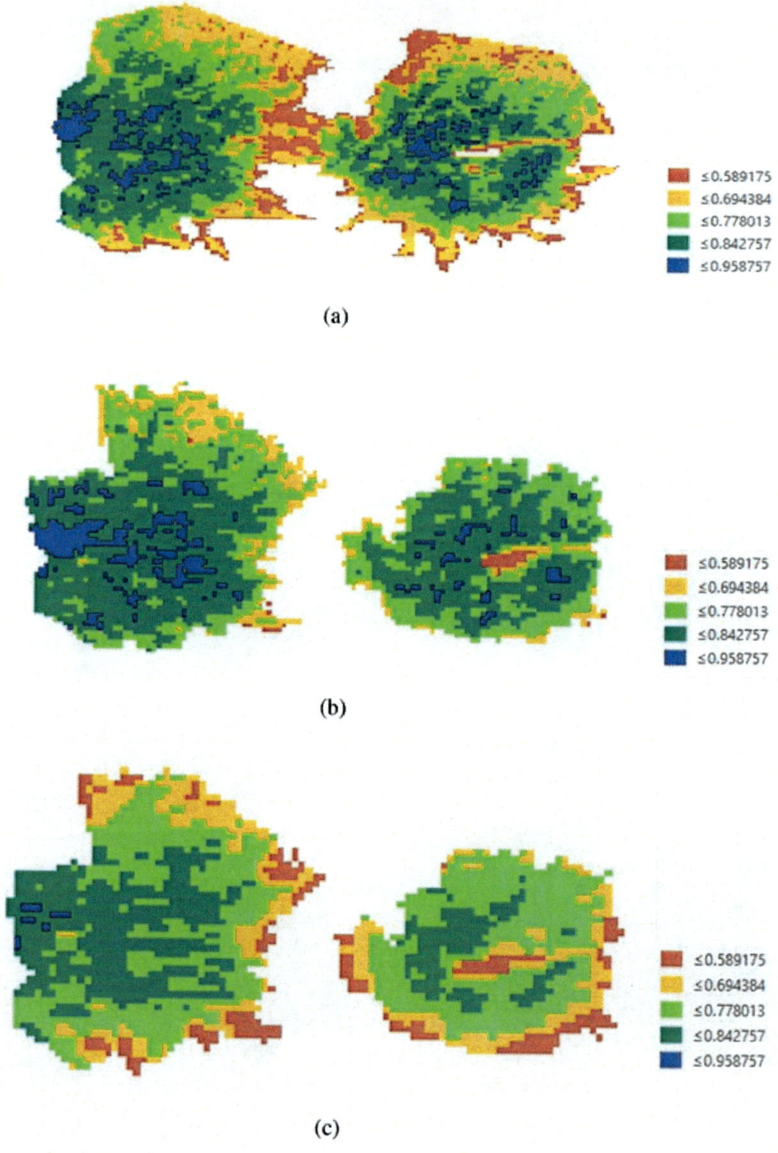

Fig. 3.16 The NDVI distribution of two individual lysimeter trees at 60 m, 90 m, and 120 m on Oct 3rd, 2019. (**a**) The NDVI distribution of individual lysimeter trees at 60 m flight height. (**b**) The NDVI distribution of individual lysimeter trees at 90 m flight height. (**c**) The NDVI distribution of individual lysimeter trees at 120 m flight height

Table 3.4 Entropy was used in the quantitative analysis and evaluation of image information, because it provided better comparison of the image details. Higher value of entropy meant more detailed information in the image

Dates	60 m	90 m	120 m
May 8th, 2019	2.0176	2.0789	1.7984
Sep 19th, 2019	4.5861	3.7598	3.1027
Oct 3rd, 2019	3.2812	3.6725	4.0022

3.5.3.3 Entropy of Individual Tree-level NDVI Image

To characterize the texture of the individual tree-level NDVI image of the lysimeter trees, the authors adopted the entropy method, which was a statistical measurement of the randomness in the image. Entropy was defined as −sum(p.*log2(p)), where p contained the normalized histogram counts returned from the MATLAB command "imhist" [32]. It was used in the quantitative analysis and evaluation of image details. Higher value of entropy meant more detailed information in the image.

Key Observation As shown in Table 3.4, the higher entropy value for lysimeter trees was obtained at 90 m on May 8th. However, the higher value entropy were at 60 m and 120 m on Sep 19th and Oct 3rd, respectively. Therefore, high spatial resolution had low impact on the UAV-based individual tree-level NDVI images. There was no significant difference for image information in range of 60–120 m of UAV flight height.

3.5.4 Conclusions and Future Work

In this study, UAV flight missions were conducted to collect multispectral imagery at different flight height (60 m, 90 m, and 120 m). Using the NDVI derived from UAV images, the authors analyzed how spatial resolution could affect the NDVI values at the individual tree level. According to the results, there was no significant difference for mean NDVI value at individual tree level for different flight heights when the data was processed appropriately. The R^2 of mean NDVI values between 90 m and 120 m was around 0.7, which proved the multispectral sensor was reliable for data collection. However, the shade around the tree canopy could be a key factor for tree canopy segmentation (Fig. 3.11), which could reduce the mean NDVI significantly.

Lower flight height (60 m) did give a higher spatial resolution image. The NDVI distribution inside the canopy was more precise than that in higher flight height. Therefore, what the average tells us can be wrong. How to use this high-resolution benefit from UAV imagery can be discussed in the near future. Variability analysis may play an important role in individual tree-level research. In the future, the authors will apply the conclusion to NDVI related research topics, such as evapotranspiration estimation, stem water potential, and yield estimation.

3.6 Chapter Summary

Compared with traditional remote sensing tools, the UAV and lightweight payloads can be more flexible and frequent in the field. UAVs can fly at a lower altitude and can take higher spatial and temporal resolution images of crops. As a low-cost scientific data collection platform, UAVs also make data acquisition relatively less expensive. While there are many advantages by using UAVs for agricultural research, the authors also discussed the challenges for UAV flight missions with different remote sensing payloads.

Based on the current research results, the authors present the following potential research directions for future investigation.

1. For future research, uncooled thermal cameras can be used to evaluate with other temperatures sensor information to acquire reliable thermal information, such as atmospherically corrected satellite images and temperature canopy sensors.
2. How to benefit the research study from high-resolution UAV imagery can be discussed in the near future.
3. Is there any optimal point where the data can be the best representation of crops or trees? As researchers try to understand the potential of the UAVs, efficient workflow, image processing methods, and better software are still under developing.

References

1. Awais, M., Li, W., Cheema, M.M., Hussain, S., Shaheen, A., Aslam, B., Liu, C., Ali, A.: Assessment of optimal flying height and timing using high-resolution unmanned aerial vehicle images in precision agriculture. Int. J. Environ. Sci. Technol. **19**, 2703–2720 (2022)
2. Berni, J.A., Zarco-Tejada, P.J., Suárez, L., Fereres, E.: Thermal and narrowband multispectral remote sensing for vegetation monitoring from an unmanned aerial vehicle. IEEE Trans. Geosci. Remote Sensing **47**(3), 722–738 (2009)
3. Brenner, C., Thiem, C.E., Wizemann, H.D., Bernhardt, M., Schulz, K.: Estimating spatially distributed turbulent heat fluxes from high-resolution thermal imagery acquired with a UAV system. Int. J. Remote Sensing **38**(8–10), 3003–3026 (2017)
4. Brown, K., Rosenberg, N.J.: A resistance model to predict evapotranspiration and its application to a sugar beet field 1. Agron. J. **65**(3), 341–347 (1973)
5. Chávez, J.L., Gowda, P., Howell, T., Neale, C., Copeland, K.: Estimating hourly crop ET using a two-source energy balance model and multispectral airborne imagery. Irrig. Sci. **28**(1), 79–91 (2009)
6. Elarab, M.: The application of unmanned aerial vehicle to precision agriculture: chlorophyll, nitrogen, and evapotranspiration estimation. Ph.D. Thesis, Utah State University (2016)
7. Ershadi, A., McCabe, M.F., Evans, J.P., Walker, J.P.: Effects of spatial aggregation on the multi-scale estimation of evapotranspiration. Remote Sensing Environ. **131**, 51–62 (2013)
8. Gade, R., Moeslund, T.B.: Thermal cameras and applications: a survey. Mach. Vis. Appl. **25**(1), 245–262 (2014)
9. Gowen, A., Tiwari, B., Cullen, P., McDonnell, K., O'Donnell, C.: Applications of thermal imaging in food quality and safety assessment. Trends Food Sci. Technol. **21**(4), 190–200 (2010)

10. Hardin, P.J., Hardin, T.J.: Small-scale remotely piloted vehicles in environmental research. Geog. Compass **4**(9), 1297–1311 (2010)
11. Hardin, P.J., Jensen, R.R.: Small-scale unmanned aerial vehicles in environmental remote sensing: challenges and opportunities. GISci. Remote Sensing **48**(1), 99–111 (2011)
12. Harwin, S., Lucieer, A.: Assessing the accuracy of georeferenced point clouds produced via multi-view stereopsis from unmanned aerial vehicle (UAV) imagery. Remote Sensing **4**(6), 1573–1599 (2012)
13. Hoffmann, H., Nieto, H., Jensen, R., Guzinski, R., Zarco-Tejada, P., Friborg, T.: Estimating evaporation with thermal UAV data and two-source energy balance models. Hydrol. Earth Syst. Sci. **20**(2), 697–713 (2016)
14. Hunsaker, D.J., Pinter, P.J., Barnes, E.M., Kimball, B.A.: Estimating cotton evapotranspiration crop coefficients with a multispectral vegetation index. Irrig. Sci. **22**(2), 95–104 (2003)
15. Jones, H.G., Serraj, R., Loveys, B.R., Xiong, L., Wheaton, A., Price, A.H.: Thermal infrared imaging of crop canopies for the remote diagnosis and quantification of plant responses to water stress in the field. Funct. Plant Biol. **36**(11), 978–989 (2009)
16. Kamble, B., Kilic, A., Hubbard, K.: Estimating crop coefficients using remote sensing-based vegetation index. Remote Sensing **5**(4), 1588–1602 (2013)
17. Laliberte, A.S., Rango, A., Herrick, J.: Unmanned aerial vehicles for rangeland mapping and monitoring: a comparison of two systems. In: Proceedings of the ASPRS Annual Conference (2007)
18. Lelong, C., Burger, P., Jubelin, G., Roux, B., Labbé, S., Baret, F.: Assessment of unmanned aerial vehicles imagery for quantitative monitoring of wheat crop in small plots. Sensors **8**(5), 3557–3585 (2008)
19. Lucieer, A., Malenovskỳ, Z., Veness, T., Wallace, L.: HyperUAS—Imaging spectroscopy from a multirotor unmanned aircraft system. J. Field Robot. **31**(4), 571–590 (2014)
20. Martinelli, F., Scalenghe, R., Davino, S., Panno, S., Scuderi, G., Ruisi, P., Villa, P., Stroppiana, D., Boschetti, M., Goulart, L.R., et al.: Advanced methods of plant disease detection. A review. Agron. Sustain. Dev. **35**(1), 1–25 (2015)
21. Monteiro, C.: Evapotranspiration estimate using energy balance two source model with UAV images: a study in vineyard. Am. J. Eng. Res. **8**(21), 60–69 (2019)
22. Montibeller, Á.G.: Estimating energy fluxes and evapotranspiration of corn and soybean with an unmanned aircraft system in Ames, Iowa. Master's Thesis, University of Northern Iowa (2017)
23. Näsi, R., Honkavaara, E., Lyytikäinen-Saarenmaa, P., Blomqvist, M., Litkey, P., Hakala, T., Viljanen, N., Kantola, T., Tanhuanpää, T., Holopainen, M.: Using UAV-based photogrammetry and hyperspectral imaging for mapping bark beetle damage at tree-level. Remote Sensing **7**(11), 15,467–15,493 (2015)
24. Nieto, H., Bellvert, J., Kustas, W.P., Alfieri, J.G., Gao, F., Prueger, J., Torres-Rua, A., Hipps, L.E., Elarab, M., Song, L.: Unmanned airborne thermal and mutilspectral imagery for estimating evapotranspiration in irrigated vineyards. In: Proceedings of the 2017 IEEE International Geoscience and Remote Sensing Symposium (IGARSS), pp. 5510–5513. IEEE, Piscataway (2017)
25. Nieto, H., Kustas, W.P., Torres-Rúa, A., Alfieri, J.G., Gao, F., Anderson, M.C., White, W.A., Song, L., del Mar Alsina, M., Prueger, J.H., et al.: Evaluation of TSEB turbulent fluxes using different methods for the retrieval of soil and canopy component temperatures from UAV thermal and multispectral imagery. Irrig. Sci. **37**(3), 389–406 (2019)
26. Niu, H., Wang, D., Chen, Y.: Estimating actual crop evapotranspiration using deep stochastic configuration networks model and UAV-based crop coefficients in a pomegranate orchard. In: Autonomous Air and Ground Sensing Systems for Agricultural Optimization and Phenotyping V. International Society for Optics and Photonics (2020)
27. Niu, H., Zhao, T., Chen, Y.: Intelligent bugs mapping and wiping (iBMW): an affordable robot-driven robot for farmers. In: IEEE International Conference on Mechatronics and Automation (ICMA), pp. 397–402. IEEE, Piscataway (2019)

28. Niu, H., Zhao, T., Wang, D., Chen, Y.: Estimating evapotranspiration with UAVs in agriculture: a review. In: Proceedings of the ASABE Annual International Meeting. American Society of Agricultural and Biological Engineers (2019)
29. Niu, H., Zhao, T., Wang, D., Chen, Y.: A UAV resolution and waveband aware path planning for onion irrigation treatments inference. In: 2019 International Conference on Unmanned Aircraft Systems (ICUAS), pp. 808–812. IEEE, Piscataway (2019)
30. Ortega-Farías, S., Ortega-Salazar, S., Poblete, T., Kilic, A., Allen, R., Poblete-Echeverría, C., Ahumada-Orellana, L., Zuñiga, M., Sepúlveda, D.: Estimation of energy balance components over a drip-irrigated olive orchard using thermal and multispectral cameras placed on a helicopter-based unmanned aerial vehicle (UAV). Remote Sensing **8**(8), 638 (2016)
31. Park, S.: Estimating plant water stress and evapotranspiration using very-high-resolution (VHR) UAV imagery. Ph.D. Thesis, University of Melbourne (2018)
32. Rafael Gonzalez, C., Woods, E.R., Eddins, L.S.: Digital image processing using MATLAB. Prentice Hall. ISBN **10**, 0130085,197 (2003)
33. dos Reis, T., Monteiro, R., Garcia, E., Albuquerque, M., Espinoza, J., Ferreira, J.: Actual evapotranspiration estimated by orbital sensors, UAV and meteorological station for vineyards in the southern Brazil. In: IV Inovagri International Meeting (2017)
34. Reyes-Gonzalez, A., Hay, C., Kjaersgaard, J., Neale, C.: Use of remote sensing to generate crop coefficient and estimate actual crop evapotranspiration. In: Proceedings of the 2015 ASABE Annual International Meeting, p. 1. American Society of Agricultural and Biological Engineers (2015)
35. Ribeiro-Gomes, K., Hernández-López, D., Ortega, J.F., Ballesteros, R., Poblete, T., Moreno, M.A.: Uncooled thermal camera calibration and optimization of the photogrammetry process for UAV applications in agriculture. Sensors **17**(10), 2173 (2017)
36. Sheng, H., Chao, H., Coopmans, C., Han, J., McKee, M., Chen, Y.: Low-cost UAV-based thermal infrared remote sensing: platform, calibration and applications. In: Proceedings of the 2010 IEEE/ASME International Conference on Mechatronics and Embedded Systems and Applications (MESA), pp. 38–43. IEEE (2010)
37. Smith, G.M., Milton, E.J.: The use of the empirical line method to calibrate remotely sensed data to reflectance. Int. J. Remote Sensing **20**(13), 2653–2662 (1999)
38. Stark, B., Chen, Y.: A framework of optimal remote sensing using small unmanned aircraft systems. In: 2016 12th IEEE/ASME International Conference on Mechatronic and Embedded Systems and Applications (MESA), pp. 1–6. IEEE, Piscataway (2016)
39. Stark, B., Zhao, T., Chen, Y.: An analysis of the effect of the bidirectional reflectance distribution function on remote sensing imagery accuracy from small unmanned aircraft systems. In: Proceedings of the 2016 International Conference on Unmanned Aircraft Systems (ICUAS), pp. 1342–1350. IEEE, Piscataway (2016)
40. Torres-Rua, A.: Vicarious calibration of sUAS microbolometer temperature imagery for estimation of radiometric land surface temperature. Sensors **17**(7), 1499 (2017)
41. Trout, T.J., Johnson, L.F., Gartung, J.: Remote sensing of canopy cover in horticultural crops. HortScience **43**(2), 333–337 (2008)
42. Turner, D., Lucieer, A., Watson, C.: An automated technique for generating georectified mosaics from ultra-high resolution unmanned aerial vehicle (UAV) imagery, based on structure from motion (SfM) point clouds. Remote Sensing **4**(5), 1392–1410 (2012)
43. Wallace, L., Lucieer, A., Watson, C., Turner, D.: Development of a UAV-LiDAR system with application to forest inventory. Remote Sensing **4**(6), 1519–1543 (2012)
44. Xia, T., Kustas, W.P., Anderson, M.C., Alfieri, J.G., Gao, F., McKee, L., Prueger, J.H., Geli, H.M., Neale, C.M., Sanchez, L., et al.: Mapping evapotranspiration with high-resolution aircraft imagery over vineyards using one-and two-source modeling schemes. Hydrol. Earth Syst. Sci. **20**(4), 1523 (2016)
45. Xiang, H., Tian, L.: Method for automatic georeferencing aerial remote sensing (RS) images from an unmanned aerial vehicle (UAV) platform. Biosyst. Eng. **108**(2), 104–113 (2011)
46. Zhang, H., Anderson, R.G., Wang, D.: Satellite-based crop coefficient and regional water use estimates for Hawaiian sugarcane. Field Crops Res. **180**, 143–154 (2015)

47. Zhao, T., Chen, Y., Ray, A., Doll, D.: Quantifying almond water stress using unmanned aerial vehicles (UAVs): correlation of stem water potential and higher order moments of non-normalized canopy distribution. In: ASME 2017 International Design Engineering Technical Conferences and Computers and Information in Engineering Conference. American Society of Mechanical Engineers (2017)
48. Zhao, T., Doll, D., Chen, Y.: Better almond water stress monitoring using fractional-order moments of non-normalized difference vegetation index. In: 2017 ASABE Annual International Meeting, p. 1. American Society of Agricultural and Biological Engineers (2017)
49. Zhao, T., Koumis, A., Niu, H., Wang, D., Chen, Y.: Onion irrigation treatment inference using a low-cost hyperspectral scanner. In: Multispectral, Hyperspectral, and Ultraspectral Remote Sensing Technology, Techniques and Applications VII. International Society for Optics and Photonics (2018)
50. Zhao, T., Niu, H., Anderson, A., Chen, Y., Viers, J.: A detailed study on accuracy of uncooled thermal cameras by exploring the data collection workflow. In: Autonomous Air and Ground Sensing Systems for Agricultural Optimization and Phenotyping III. International Society for Optics and Photonics (2018)
51. Zhao, T., Niu, H., de la Rosa, E., Doll, D., Wang, D., Chen, Y.: Tree canopy differentiation using instance-aware semantic segmentation. In: 2018 ASABE Annual International Meeting. American Society of Agricultural and Biological Engineers (2018)
52. Zhao, T., Stark, B., Chen, Y., Ray, A.L., Doll, D.: A detailed field study of direct correlations between ground truth crop water stress and normalized difference vegetation index (NDVI) from small unmanned aerial system (sUAS). In: 2015 International Conference on Unmanned Aircraft Systems (ICUAS), pp. 520–525. IEEE, Piscataway (2015)
53. Zhao, T., Wang, Z., Yang, Q., Chen, Y.: Melon yield prediction using small unmanned aerial vehicles. In: Autonomous Air and Ground Sensing Systems for Agricultural Optimization and Phenotyping II, vol. 10218, p. 1021808. International Society for Optics and Photonics (2017)
54. Zhao, T., Yang, Y., Niu, H., Wang, D., Chen, Y.: Comparing U-Net convolutional network with mask R-CNN in the performances of pomegranate tree canopy segmentation. In: Multispectral, Hyperspectral, and Ultraspectral Remote Sensing Technology, Techniques and Applications VII, vol. 10780, p. 107801J. International Society for Optics and Photonics (2018)

Chapter 4
The Edge-AI Sensors and Internet of Living Things (IoLT)

Check for updates

Abstract This chapter delves into the realm of Edge-AI sensors and their intersection with the Internet of Living Things (IoLT), offering a comprehensive exploration of proximate sensors and their applications. The chapter commences with an introduction, setting the stage for the exploration of cutting-edge technologies that bridge the gap between artificial intelligence and the sensory capabilities of the Internet of Things (IoT). A detailed examination of proximate sensors unfolds, with a focus on innovative devices such as spectrometers, pocket-sized spectrometers, and microwave radio frequency 3D sensors. Each sensor type is dissected to unveil its unique functionalities and potential applications in the context of Edge-AI and IoLT. The chapter presents a compelling case study that demonstrates the practical implications of Edge-AI sensors in agriculture. The study investigates onion irrigation treatment inference using a low-cost Edge-AI sensor, offering insights into the potential for efficient and cost-effective digital agriculture practices. The case study includes a thorough exploration of materials, methods, results, and discussion, culminating in conclusions and suggestions for future research directions.

4.1 Introduction

The term "Internet of Things" was first created by Kevin Ashton of Procter and Gamble, later MIT's Auto-ID Center, in 1999 [1]. The Internet of Things (IoT) can describe physical objects that are embedded with sensors, software, and other technologies, which connect and exchange information with other equipment or systems over the Internet or other communication networks [13, 19, 27]. This research topic has been gaining interest because of the convergence of multiple technologies, such as ubiquitous computing, commodity sensors, embedded systems, and machine learning. There are extensive applications for IoT devices, such as agriculture, medical, and healthcare [5, 9] and manufacturing [30]. For example, many researchers have been working on the application of the IoT technology in agriculture [2, 15, 23]. In [31], Zhao et al. selected mobile wireless communication technology to achieve greenhouse monitoring with IoT technology.

H. Niu, Y. Chen, *Smart Big Data in Digital Agriculture Applications*, Agriculture Automation and Control, https://doi.org/10.1007/978-3-031-52645-9_4

Research results showed that the greenhouse monitoring system based on IoT technology could automatically control the environmental temperature and humidity factors. In [12], Hamad et al. emphasized the role of smartphones for collecting agricultural parameters and its advantages in the field of agriculture. Using the smartphone enabled the farmers to have a positive impact on communication with producer's network and improved the awareness of farming community.

Considering the authors' research experience in agriculture, they proposed the concept of "Internet of Living Things (IoLT)" in this chapter. In the dictionary, a "living thing" pertains to any organism or a life form that possesses or shows the characteristics of life or being alive. The fundamental characteristics are (1) having an organized structure, (2) requiring energy, (3) responding to stimuli and adapting to environmental changes, and (4) being capable of reproduction, growth, movement, metabolism, and death. For most of the agricultural applications of IoT, it actually can also be considered as the IoLT. Researchers collect data of crops or trees, such as tree canopy temperature, rainfall, humidity, wind speed, soil moisture, stem water potential, etc. After the analysis of recorded data, stakeholders or researchers can help make better decisions to increase the farm productivity, improve the sustainability, and reduce the costs.

The proximate sensors play a key role in IoLT. Therefore, in the rest of this chapter, the authors introduced several sensors they commonly used in the research work.

4.2 Proximate Sensors

4.2.1 The Spectrometer

The spectrometer EPP2000-VIS-50 was used for the onion irrigation treatment inference research. The StellarNet EPP2000 Spectrometers (StellarNet, Inc., Tampa, FL, USA) are miniature fiber optic instruments for ultraviolet (UV), visible (VIS), and near-infrared (NIR) measurements in 190–1700 nm ranges (Fig. 4.1). The technical specifications of EPP2000-VIS-50 are listed in Table 4.1 for reference.

4.2.2 A Pocket-Sized Spectrometer

Recently, a light and small near-infrared spectrometer, Scio (Consumer Physics, Israel), was released in the market (Fig. 4.2). They used advances in micro-optical technology to miniaturize the traditional near-infrared spectrometer. The optical head is just a few millimeters in size and can provide sensitivity and accuracy levels on par with the best bench spectrometers. The low power consumption and zero warm-up time make it highly responsive and extremely efficient, which allows

Fig. 4.1 The spectrometer EPP2000-VIS-50. The StellarNet EPP2000 Spectrometers are miniature fiber optic instruments for ultraviolet (UV), visible (VIS), and near-infrared (NIR) measurements in 190–1700 nm ranges

Table 4.1 The specifications of spectrometer EPP2000-VIS-50

Specifications	
Wavelength range	190–1700 nm
Base unit size	15.5 × 9.5 × 4.0 cm
Base unit mass	500 g
Detector	2048 pixel, 14 × 200 μm
Linear range	0–2.1 absorbance units (<0.5%)
Exposure range	4 ms to 60 s
Integration time range	4–6500 ms
Wavelength accuracy	<0.25 nm
Wavelength repeatability	<0.05 nm
Wavelength stability	<0.001 nm per °C

it to perform hundreds of samplings from a small rechargeable battery. The Scio spectrometer works in NIR at wavelengths of 700–1100 nm. It is so small that it can even be integrated into smart phone as an Edge-AI device. This system is a low-cost module, with a price less than $300. The technical specifications of Scio are listed in Table 4.2 for reference.

4.2.3 A Microwave Radio Frequency 3D Sensor

The reflectance of the environment being sensed was collected with a radio frequency 3D sensor called Walabot Developer (Fig. 4.3) (Vayyar Imaging Ltd, Yehud-Monosson, Israel). The frequency range was 3.3–10 GHz for the US/FCC model and 6.3–8 GHz for the EU/CE model. The average transmitted energy of both models was below 41 dBm/MHz and did not constitute health concerns for the user, nor did it impact the tested tissues. Based on the technical specifications,

Fig. 4.2 A light and small near-infrared spectrometer, Scio. The low power consumption and zero warm up time make it highly responsive and extremely efficient, which allows it to perform hundreds of samplings from a small rechargeable battery. The Scio spectrometer works in NIR at wavelengths of 700–1100 nm. It can also be integrated into smart phone as an Edge-AI device

Table 4.2 The specifications of spectrometer Scio

Technical specifications	
Wavelength range	700–1100 nm
Dimensions Handheld Device	Cover: 67.7 × 40.2 × 18.8 mm
Weight handheld device	35 g
Standalone sensor module	27.5 × 9.5 × 3.15 mm
Operational distance	Contact to 2 cm
Typical scan time	2–5 s
Connectivity	Online and offline scanning supported
Temperature range—operation	4–35 °C

the Walabot could sense the environment by transmitting, receiving, and recording signals from multiple antennas. Recordings of multiple transmit-receive antenna pairs were analyzed to build a 3D image of the examined environment. Changes in the environment were characterized by analyzing sequences of images. Then, researchers could detect changes in the environment by analyzing the sequences of images.

The Walabot used an antenna array to send radio frequency to the area in front of it and then captured the returning signals (reflectance or response). The citation signals were produced, and the reflectance was then recorded by integrated circuits (Fig. 4.4). The reflectance data were preprocessed by MATLAB 2020a. For example, records of the different reflectance strengths could be visualized in a graph (Fig. 4.5). The y-axis denoted the radio frequency reflectance from a walnut leaf. The x-axis meant the time. The research hypothesis was that the radio frequency reflectance of walnut leaves would be significantly different toward root-lesion nematode infection levels. Then, the spectral curve obtained from the leaves of walnut trees infected with nematodes could show the characteristic pattern of walnut under nematode stress. For more details on this research work, please refer to Chap. 7.

Fig. 4.3 A radio frequency 3D sensor, Walabot Developer (Front and back view). The frequency range was 3.3–10 GHz for the US/FCC model and 6.3–8 GHz for the EU/CE mode. The average transmitted energy of both models was below 41 dBm/MHz and did not constitute health concerns for the user, nor did it impact the tested tissues

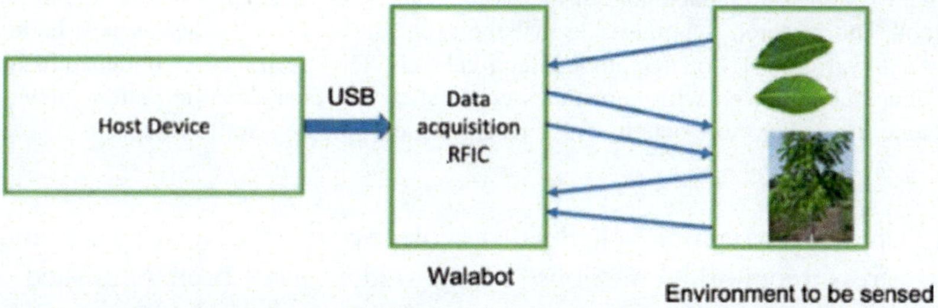

Fig. 4.4 The principle of Walabot. The sensor uses an antenna array to send radio frequency to the area in front of it, e.g. walnut leaves, and then captures the returning signals (reflectance or response). The citation signals are produced, and the reflectance is then recorded by the radio frequency integrated circuits (RFIC)

4.3 Case Study: Onion Irrigation Treatment Inference Using a Low-Cost Edge-AI Sensor

4.3.1 Introduction

Onions are produced and consumed throughout the world. It is worldwide used in different countries and cultures during all seasons in a year [4]. California produces the most onion in the USA. In 2015, it produced around one-third of the total onion crop in the USA. It is the only state that can produce spring and summer-harvested onions [18]. Onions are shallow-rooted crop, and most of the roots can be found in the top 0.18 m of soil [7]. This makes it hard for onions to obtain enough soil water. Therefore, lighter and more frequent irrigation are recommended in onion cultural practices [6]. On the other hand, experimental results showed that water stress could also cause reduction in the yield at any growing stages [26].

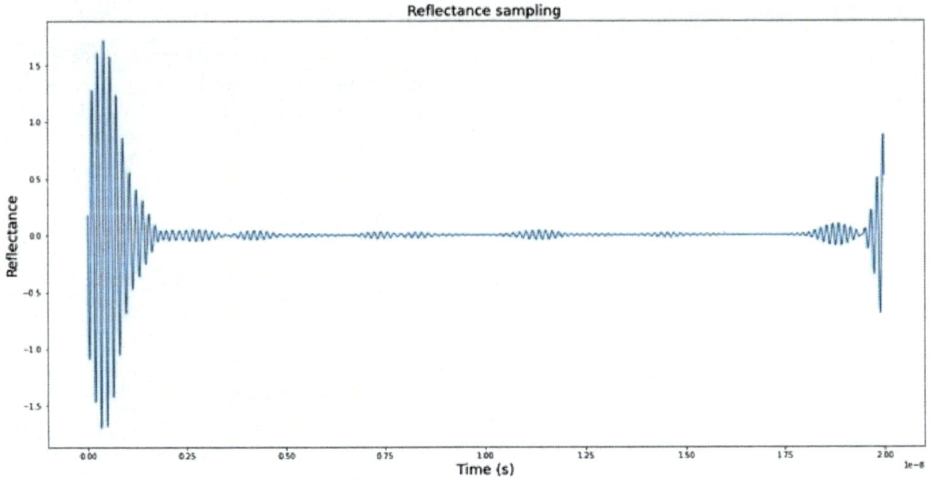

Fig. 4.5 The measurement signal of the radio frequencies reflectance of the walnut leaves using Walabot. The research hypothesis is that the reflectance of walnut leaves will be significantly different toward root-lesion nematode infection levels. The spectral curve obtained from the leaves of walnut trees infected with nematodes could show the characteristic pattern of walnut under nematode stress. For more details of this research work, please refer to Chap. 7

To optimize irrigation schedule, it is necessary to have accurate and reliable water stress monitoring methods. Many studies have been published on water stress detection using remote sensing, a real-time and nondestructive method [11]. Near-infrared cameras were used to detect water stress of almond trees [32–37], where new types of spectral features were proposed to predict stem water potential. Hyperspectral sensors were also applied for water stress monitoring in apple trees [17], vineyard [10, 20], etc. For onions, studies of remote sensing were conducted on yield and biomass prediction using the spectroradiometer [21], biomass monitoring using unmanned aerial vehicles and RGB cameras [3], detection of diseased onion tissues [29], and quality inspections [14, 28]. However, to date, no one has studied irrigation treatment inference using the spectrometer. Furthermore, Scio, as a low-cost portable, light spectrometer, was also evaluated to infer irrigation treatments in onions for the first time.

4.3.2 *Materials and Methods*

4.3.2.1 Onion Study Site

The study field was in USDA-ARS, San Joaquin Valley Agricultural Sciences Center (36.59 °N, 119.51 °W), Parlier, CA, USA (Fig. 4.6). Since 2016, an onion test field had been set up for research of biomass soil amendments and deficit irrigation. There were 36 plots in total. Each plot had two 40 inches wide beds

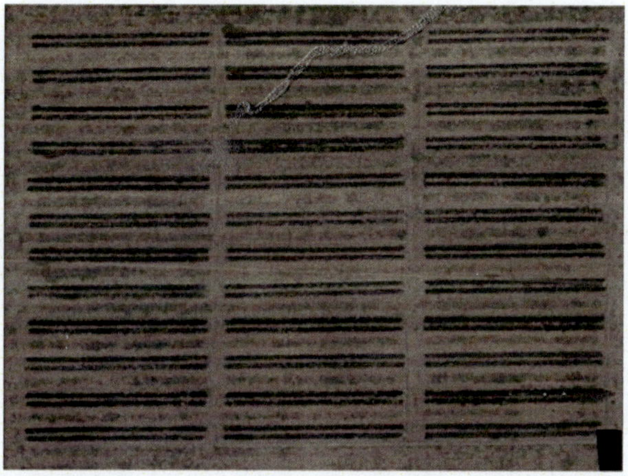

Fig. 4.6 The onion field (image taken by the "Hover" UAV platform). The study field was in USDA-ARS, San Joaquin Valley Agricultural Sciences Center (36.59 °N, 119.51 °W), Parlier, CA, USA. Since 2016, an onion test filed had been set up for research of biomass soil amendments and deficit irrigation. There were 36 plots in total. Each plot had two 40 inches wide beds by 20 meters long. There were three irrigation treatment levels, High, Medium, and Low

by 20 meters long. There were three levels of irrigation treatment, High, Medium, and Low, and four soil amendments, Biochar, Check, Biochar + Compost, and Biochar + Compost + Sulfur. There were three replicate plots for each treatment combination.

4.3.2.2 The Spectrometer Scio

Most recently, a light and small hyperspectral scanner called Scio (Consumer Physics, Israel) was released in the market. As a complete system, it included a spectrometer, a light source, and optimized algorithms in the cloud. The Scio spectrometer worked in NIR at wavelengths of 700–1100 nm (Fig. 4.7). It could also be integrated into the smart phone. This system was a low-cost module, with the price less than $300.

As a reference, the author also used a traditional handheld spectrometer EPP2000-VIS-50. Its wavelength sensitivity was from 190 nm to 1700 nm, with the wavelength accuracy of 0.25 nm.

4.3.2.3 Field Measurement Collection

During the growing season, onions under different treatment were sampled once a month. The field measurements using the spectrometers were coordinated with these physiological measurements, including shoot length, root length, number of leaves, fresh weight, dry weight, and bulb diameter. There were three onion samples

Fig. 4.7 The NIR reflectance measurement of SCIO using smart phone

collected for each plot. For each onion sample, three measurements were made using both the EPP2000-VIS-50 (Fig. 4.8) and Scio at the same time in the field to reduce the likelihood of errors or anomalous results. In general, the author had 81 measurements for 27 onion samplings.

"SCIO" was an active sensing platform and it provided calibration case. The sensor was first calibrated using the white panel in the case. The measurement using "SCIO" required the distance between onion leaves and the scanner as small as possible to minimize the disturbance of sunlight source. On the other hand, EPP2000-VIS-50 was a passive sensing platform. The author needed to take the measurements of white panel and dark panel for calibration before measuring the onion leaves.

4.3.2.4 Principal Component Analysis

Both principal component analysis (PCA) and linear discriminant analysis (LDA) were methods for reducing dimensionality of a dataset to increase accuracy, speed up processing time, and generate visualization.

PCA is a linear transformation that rotates the axes of the data along the direction that maximizes its variance, allowing data to be projected onto a lower dimensional subspace [16]. These new axes, or "loadings," are found by calculating the eigenvectors W of the covariance matrix of data, where X is an $M \times N$ matrix representing M samples of size N:

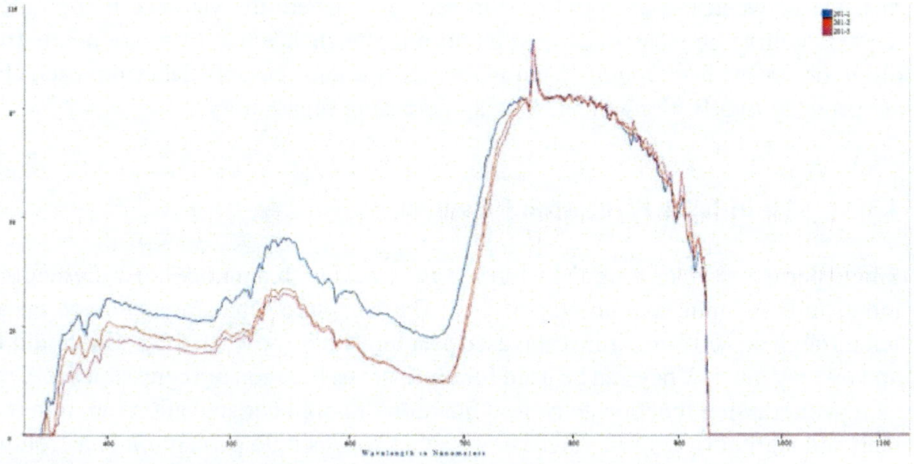

Fig. 4.8 The reflectance measurement of EPP2000-VIS-50. Each onion sampling was measured three times to reduce the likelihood of errors or anomalous results

$$X^T X = W\lambda. \tag{4.1}$$

The eigenvalues λ represented how "important" each loading was in transforming the data or how much variance the projection onto this axis contributed. As the loadings (and λ) were sorted in descending order, W could be truncated to r columns, which could then be used to project data along r dimensions, preserving the dimensions that contributed most to the variance of distribution. W was often obtained with singular value decomposition (SVD) instead of performing the eigen decomposition of $X^T X$, as it was more computationally efficient.

4.3.2.5 Linear Discriminant Analysis

LDA reduces dimensionality of data by finding new axes to project it onto that maximizes the separability between classes [8]. It does this by maximizing the distance between means of classes relative to some center point for all classes while minimizing the variance, or scatter, within each category (Eq. (4.2)). In the following equation, C is the number of classes, N_i is the size of class i, μ is the mean of all datapoints, μ_i is the mean of class i, and x_j is the jth datapoint in class i

$$\frac{\sum_{i=1}^{C} N_i(\mu_i - \mu)(\mu_i - \mu)^T}{\sum_{i=1}^{C}\sum_{j=1}^{N_i}(x_j - \mu_i)(x_j - \mu_i)^T}. \tag{4.2}$$

The optimized solution contains eigenvectors that is descending order of their eigenvalues, which can be used to reduce the dataset similar to PCA. Optimizing for both within- and between-class scatter is important because only maximizing the distance between means could lead to scenarios where the variance is high along the axis with large mean distances, increasing the chance that there are points from different classes overlapping. Minimizing the variance ensures data from each class are grouped tightly along the new axis, increasing separability.

4.3.2.6 Multi-layer Perceptron Classifier

Single perceptrons, or artificial neurons, are nodes with a number of weighted data inputs, a bias input, and an output [25]. The weighted inputs are summed up and fed through an activation function, used as a threshold for when the node should fire and by how much. They can be used for binary classification or regression problems, using either a step function activation function or nonlinear activation function (e.g., tanh and sigmoid), respectively. The function for a single perceptron is as follows, where s is the activation function, W is the input weights, and b is the bias

$$f(x) = s(b + Wx). \tag{4.3}$$

Single perceptrons cannot be used for many complex prediction tasks because they can only predict nonlinear patterns [22]. Multi-Layer Perceptrons (MLPs) overcome this by constructing networks out of multiple perceptrons. An MLP is a supervised learning system consisting of an input layer, N number of hidden layers, and an output layer. Nonlinear activation functions used in MLP's introduce nonlinearities into the model, allowing it to make predictions on complex, nonlinear datasets, such as hyperspectral readings. MLPs are trained using a process called backpropagation, which updates the network's weights with respect to the error between its current output and the expected result.

4.3.3 Results and Discussion

To prepare the dataset of EPP2000-VIS-50, each 1675-dimensional reading was loaded into a vector, with readings of the same plant being treated as different data points. "SCIO" measurements were obtained with the help of smart phone application, and each reading was of 1060 dimensions. For each test iteration, data were split into a 75% for training and 25% for testing. All data were normalized with scikit-learn's $normalize()$ function, while the MLP classifier required normalization with the $StandardScalar()$ module [24].

The evaluation stage was broken up into three steps: data preparation, classifier evaluation, and parameter grid search on the best performing classifiers. The classifier performance was ranked by the percentage of correctly predicted labels in the test dataset, averaged over ten iterations with 75%/25% train/test splits as described above. In the classifier evaluation stage, a number of classifiers were being applied with and without PCA and LDA dimensionality reduction.

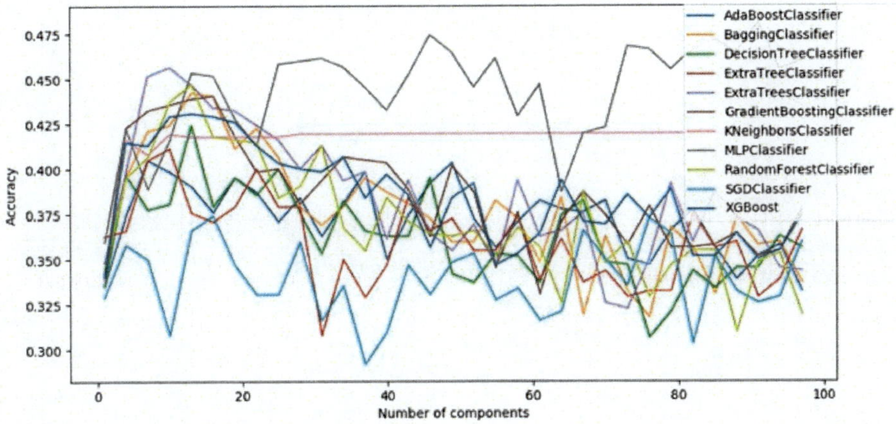

Fig. 4.9 Classifier accuracy comparison on PCA reduced "SCIO" dataset

Table 4.3 Top five performing classifiers using PCA and MLP and their grid search parameters

Accuracy	PCA components	Hidden layer sizes	Activation	Solver	Alpha	Batch size	Max iterations
0.53064	20	100	tanh	adam	0.001	25	500
0.52903	20	50	relu	lbfgs	0.1	50	500
0.52741	20	200	tanh	adam	1e-05	25	500
0.52741	20	400	relu	adam	0.01	100	500
0.52580	20	50	logistic	adam	1e-05	25	1000

4.3.3.1 Results Using PCA-Based Classifiers

Figure 4.9 shows the performance of several scikit-learn classifiers (and XGBoost) ranked against each other in terms of label prediction accuracy for the "SCIO" dataset (reduced with the PCA from 3 to 99 components). Most classifiers performed very poorly, with only the MLP classifier nearly breaking 50% accuracy, hitting a high score of 48.1% at 90 components, with the rest barely beating random guessing (33% for 3 labels). After MLP was determined to perform the best with PCA reduced data, the author performed parameter grid search to optimize its results, iterating through the parameters listed below, achieving a top accuracy of 53.1% (top five results and parameters are shown in Table 4.3).

- PCA components: 10, 20, 40, 80, and 160
- Hidden layer sizes: 25, 50, 100, 200, and 400
- Activation: relu, logistic, and tanh
- Solver: lbfgs and adam
- Alpha: 0.00001, 0.0001, 0.001, 0.01, and0.1
- Batch size: 200, 100, 50, and 25
- Max iterations: 200, 500, and 1000

prediction accuracy of MLP from 48.1% (with 90 components) to 77.4% (with 30 components) for PCA and 39.5% (with 6 components) to 58.5% (with 42 components) for LDA.

In the future, the authors will collect more measurements in different growing stages of onions to make sure the model is robust to the different growth stages. As for "SCIO," new machine learning algorithms should be explored to see if better algorithms can help improve the prediction accuracy.

4.4 Chapter Summary

In this chapter, the authors proposed the concept of "Internet of Living Things (IoLT)." A "living thing" pertains to any organism or a life form that possesses or shows the characteristics of life or being alive. The fundamental characteristics are (1) having an organized structure, (2) requiring energy, (3) responding to stimuli and adapting to environmental changes, and (4) being capable of reproduction, growth, movement, metabolism, and death. For most of the agricultural applications of IoT, it actually can also be considered as the IoLT. Researchers collect data of crops or trees, such as tree canopy temperature, rainfall, humidity, wind speed, soil moisture, stem water potential, etc. After the analysis of recorded data, stakeholders or researchers can help make better decisions to increase the farm productivity, improve the sustainability, and reduce the costs. For the future research, new sensors can be designed for specific agricultural applications. How to extract the most significant data using the ML algorithms can be an important research topic.

References

1. Ashton, K., et al.: That 'Internet of Things' thing. RFID J. **22**(7), 97–114 (2009)
2. Balducci, F., Impedovo, D., Pirlo, G.: Machine learning applications on agricultural datasets for smart farm enhancement. Machines **6**(3), 38 (2018)
3. Ballesteros, R., Ortega, J.F., Hernandez, D., Moreno, M.A.: Onion biomass monitoring using UAV-based RGB imaging. Precis. Agric. **19**(5), 840–857 (2018)
4. Córcoles, J.I., Ortega, J.F., Hernández, D., Moreno, M.A.: Estimation of leaf area index in onion (Allium cepa L.) using an unmanned aerial vehicle. Biosyst. Eng. **115**(1), 31–42 (2013)
5. Da Costa, C.A., Pasluosta, C.F., Eskofier, B., Da Silva, D.B., da Rosa Righi, R.: Internet of health things: toward intelligent vital signs monitoring in hospital wards. Artif. Intell. Med. **89**, 61–69 (2018)
6. Doneen, L.D., MacGillivray, J.H.: Suggestions on irrigating commercial truck crops. University of California, College of Agriculture, Agricultural Experiment Station (1943)
7. Drinkwater, W., Janes, B.: Effects of irrigation and soil moisture on maturity, yield and storage of two onion hybrids. In: Proceedings of the American Society for Horticultural Science, vol. 66, pp. 267–278 (1955)
8. Fisher, R.A.: The use of multiple measurements in taxonomic problems. Ann. Eugen. **7**(2), 179–188 (1936)

9. Gatouillat, A., Badr, Y., Massot, B., Sejdić, E.: Internet of medical things: a review of recent contributions dealing with cyber-physical systems in medicine. IEEE Internet Things J. **5**(5), 3810–3822 (2018)
10. González-Fernández, A.B., Rodríguez-Pérez, J.R., Marcelo, V., Valenciano, J.B.: Using field spectrometry and a plant probe accessory to determine leaf water content in commercial vineyards. Agric. Water Manag. **156**, 43–50 (2015)
11. Govender, M., Govender, P., Weiersbye, I., Witkowski, E., Ahmed, F.: Review of commonly used remote sensing and ground-based technologies to measure plant water stress. Water SA **35**(5), 741–752 (2009)
12. Hamad, M.A., Eltahir, M.E.S., Ali, A.E.M., Hamdan, A.M.: Efficiency of using smart-mobile phones in accessing agricultural information by smallholder farmers in North Kordofan–Sudan. Available at SSRN 3240758 (2018)
13. Hassan, W.H., et al.: Current research on Internet of Things (IoT) security: a survey. Comput. Netw. **148**, 283–294 (2019)
14. Islam, M.N., Nielsen, G., Stærke, S., Kjær, A., Jørgensen, B., Edelenbos, M.: Novel non-destructive quality assessment techniques of onion bulbs: a comparative study. J. Food Sci. Technol. **55**, 3314–3324 (2018)
15. Jayaraman, P.P., Yavari, A., Georgakopoulos, D., Morshed, A., Zaslavsky, A.: Internet of Things platform for smart farming: experiences and lessons learnt. Sensors **16**(11), 1884 (2016)
16. Jolliffe, I.: Principal Component Analysis. Springer, Berlin (2011)
17. Kim, Y., Glenn, D.M., Park, J., Ngugi, H.K., Lehman, B.L.: Hyperspectral image analysis for water stress detection of apple trees. Comput. Electron. Agric. **77**(2), 155–160 (2011)
18. Lazicki, P., Geisseler, D., Horwath, W.R.: Onion production in California. Accessed 16 Aug 2018
19. Madakam, S., Lake, V., Lake, V., Lake, V., et al.: Internet of Things (IoT): a literature review. J. Comput. Commun. **3**(05), 164 (2015)
20. Maimaitiyiming, M., Ghulam, A., Bozzolo, A., Wilkins, J.L., Kwasniewski, M.T.: Early detection of plant physiological responses to different levels of water stress using reflectance spectroscopy. Remote Sensing **9**(7), 745 (2017)
21. Marino, S., Basso, B., Leone, A., Alvino, A.: Agronomic traits and vegetation indices of two onion hybrids. Sci. Hortic. **155**, 56–64 (2013)
22. Minsky, M., Papert, S.: Perceptrons. MIT Press, Cambridge (1969)
23. Paustian, M., Theuvsen, L.: Adoption of precision agriculture technologies by German crop farmers. Precis. Agric. **18**(5), 701–716 (2017)
24. Pedregosa, F., Varoquaux, G., Gramfort, A., Michel, V., Thirion, B., Grisel, O., Blondel, M., Prettenhofer, P., Weiss, R., Dubourg, V., et al.: Scikit-learn: machine learning in Python. J. Mach. Learn. Res. **12**(Oct), 2825–2830 (2011)
25. Rosenblatt, F.: The perceptron: a probabilistic model for information storage and organization in the brain. Psychol. Rev. **65**(6), 386 (1958)
26. Singh, R., Alderfer, R.: Effects of soil-moisture stress at different periods of growth of some vegetable crops. Soil Sci. **101**(1), 69–80 (1966)
27. Union, I.T.: Internet of Things global standards initiative (2012)
28. Wang, W., Li, C.: A multimodal machine vision system for quality inspection of onions. J. Food Eng. **166**, 291–301 (2015)
29. Wang, W., Li, C., Gitaitis, R.D.: Optical properties of healthy and diseased onion tissues in the visible and near-infrared spectral region. Trans. ASABE **57**(6), 1771–1782 (2014)
30. Yang, C., Shen, W., Wang, X.: The Internet of Things in manufacturing: key issues and potential applications. IEEE Syst. Man Cybern. Mag. **4**(1), 6–15 (2018)
31. Zhao, J., Zhang, J., Feng, Y., Guo, J.: The study and application of the IoT technology in agriculture. In: 2010 3rd International Conference on Computer Science and Information Technology, vol. 2, pp. 462–465. IEEE, Piscataway (2010)
32. Zhao, T., Chen, Y., Ray, A., Doll, D.: Quantifying almond water stress using unmanned aerial vehicles (UAVs): correlation of stem water potential and higher order moments of non-normalized canopy distribution. In: ASME 2017 International Design Engineering Technical

Conferences and Computers and Information in Engineering Conference. American Society of Mechanical Engineers (2017)
33. Zhao, T., Doll, D., Chen, Y.: Better almond water stress monitoring using fractional-order moments of non-normalized difference vegetation index. In: 2017 ASABE Annual International Meeting, p. 1. American Society of Agricultural and Biological Engineers (2017)
34. Zhao, T., Doll, D., Wang, D., Chen, Y.: A new framework for UAV-based remote sensing data processing and its application in almond water stress quantification. In: 2017 International Conference on Unmanned Aircraft Systems (ICUAS), pp. 1794–1799. IEEE, Piscataway (2017)
35. Zhao, T., Stark, B., Chen, Y., Ray, A., Doll, D.: More reliable crop water stress quantification using small unmanned aerial systems (sUAS). IFAC-PapersOnLine **49**(16), 409–414 (2016)
36. Zhao, T., Stark, B., Chen, Y., Ray, A.L., Doll, D.: A detailed field study of direct correlations between ground truth crop water stress and normalized difference vegetation index (NDVI) from small unmanned aerial system (sUAS). In: 2015 International Conference on Unmanned Aircraft Systems (ICUAS), pp. 520–525. IEEE, Piscataway (2015)
37. Zhao, T., Stark, B., Chen, Y., Ray, A.L., Doll, D.: Challenges in water stress quantification using small unmanned aerial system (sUAS): lessons from a growing season of almond. J. Intell. Robot. Syst. **88**(2–4), 721–735 (2017)

Chapter 5
The Unmanned Ground Vehicles (UGVs) for Digital Agriculture

Abstract This chapter explores the integration of Unmanned Ground Vehicles (UGVs) in the realm of digital agriculture, offering a detailed examination of their role as data acquisition platforms. The chapter begins with an introduction, framing the discussion within the broader context of leveraging UGVs to enhance agricultural practices. An in-depth exploration follows, highlighting the UGV's significance as a versatile and accessible data acquisition tool. Fundamental research questions are addressed, emphasizing the low entry barriers associated with UGVs. The chapter delves into the application of cognitive algorithms, particularly those rooted in deep learning, and investigates the swarming mechanism of UGVs for collaborative and efficient data gathering. A compelling case study is presented that details the construction of a UGV platform for agricultural research using a low-cost toy vehicle. The study provides insights into the practical implementation of UGVs in the agricultural domain, emphasizing the feasibility of adopting cost-effective solutions for digital agriculture. The chapter concludes with a comprehensive summary, encapsulating key insights into the role of UGVs in digital agriculture. References are provided to readers interested in further exploration of the technologies, algorithms, and case studies discussed in Chap. 5, ensuring a scholarly foundation for the content presented.

5.1 Introduction

An unmanned ground vehicle (UGV) is an autonomous vehicle, capable of not only driving without a human driver but can also include artificial intelligence to conduct search, proximate sensing, and mapping and tracking mission without the need of human's control [34]. Current UGVs come in a variety of forms and capabilities, from full-sized tractor capable of replicating manned operations to miniature systems for agricultural environments [28]. Industry experts predicted a significant increase in UGV deployment, especially within the agricultural industry. This exponential increase is expected from two major developments: the increase in the availability of advanced sensor technology [35] and the cognitive applications that allow the commercial use of UGVs [17]. Advanced sensing technology allows

H. Niu, Y. Chen, *Smart Big Data in Digital Agriculture Applications*, Agriculture Automation and Control, https://doi.org/10.1007/978-3-031-52645-9_5

cameras to be automated, leading the way for UGVs to further reduce labor costs in agriculture. According to a robotic research report, agricultural robots can become a 45 billion industry by 2038 [15].

UGVs, equipped with sensors, have been used in several agricultural sensing applications for collecting data. For example, the CGMD302 spectrometer can be used to measure the leaf area index (LAI) for monitoring the rice growth conditions [29]. Equipped with pan-tilt-zoom (PTZ) camera and stereo cameras, Husky UGV from Clearpath Robotics can be used for optimized harvesting by isolating low yielding crops [1]. Velodyne VLP-16 LiDAR, equipped with the IG52-DB4 UGV from Superdroid Robots, can be used for 3D mapping of plants [30]. Furthermore, equipped with a 2D laser rangefinder, a Kinect v2 camera, and a set of robot navigation sensors, a commercial off-the-shelf mobile robot Summit XL HL can be used to assess the canopy dimensions of vines and estimate the yield [22]. Pest population distribution and irrigation non-uniformity mapping are still new areas for UGVs, but they have more exact requirements than other sensing applications. For example, real-time management of irrigation systems requires more and more precise information on water, soil, and plant conditions than most surveillance applications. Most current UGV sensing applications use expensive UGVs and cameras [1].

Pests are generally considered the most economically important factor in agriculture due to their spatial coverage [31]. Early detection of the distribution of the pest population in agricultural systems is critical to allow timely interventions and reduce crop yield losses [23]. For example, estimates are that approximately 10.8% of global crop losses are due to pests [32]. Each year in California, pest eradication is not always available, due to either the rapidly expanding distribution of the pest or constraints regarding the tools available to attempt eradication. Pests can cause significant damage to the agricultural industry and natural environments [26].

The irrigation systems have been designed to apply water uniformly in the irrigation area. Uniformity of soil moisture under the sprinkler irrigation system is an important parameter for crop quality and yield. Non-uniformity irrigation can cause less yields than expected [18]. Smart irrigation management will enable growers to irrigate ideally in agricultural industry. For example, research in almond irrigation management has been accelerated to achieve sustainability and marketability since 2010 in California. There are around 7600 almond farms in California, which makes up almost 100% of commercial almond production in the USA and about 80% of the global supply. Since there is limited water in CA, it is important to move the almond industry forward toward sustainable water management. The Sustainable Ground Management Act (SGMA) defined a new industry goal for almond water use by 2025 to "Reduce the amount of water used to grow a pound of almonds by 20%." Smart irrigation management will allow growers to ideally irrigate almonds and keep CA as a global leader in sustainable almond production.

Current state-of-the-art methods for identifying pest population distribution and non-uniformity irrigation system require farmers or growers to employ field scouts to identify visible symptoms or rely on remote sensing images from satellites or UAVs to identify crops [4]. Usually, when visual symptoms occur, the damage may already have occurred. Early detection of crop pest population distribution and irrigation non-uniformity mapping will directly or indirectly contribute to agricultural, environmental, and economic sustainability due to rational irrigation water usage, minimized application of fertilizers and pesticides, and improve yield estimation.

This chapter will first focus on how to build a low-cost UGV platform and its application for pest population distribution and irrigation non-uniformity mapping. The proposed fundamental research can also be used for other digital agriculture applications, such as leak detection. UGV, as a data acquisition platform, will definitely play a key role in these agricultural applications.

5.2 UGV as Data Acquisition Platform

Over the past decade, UGVs have shown great promise in their future role as a primary tool for collecting critical information for digital agriculture [40]. Commercial UGVs have the potential to deal with rising demand to reduce labor costs in the world. Whether used as a smart farmer's companion or a co-ecologist, cognitive swarming through UGVs with associated system applications is rapidly increasing in number and sophistication. We are limited only by our imagination for potential future applications, much like the dawn of personal computing.

The proposed UGVs in this chapter can run more than 6 hours in the field and have the ability to capture remarkably high-resolution thermal, near-infrared (NIR), red–green–blue (RGB) standard video and imagery, and LIDAR for 3D mapping. These data can then be used for a variety of timely decisions, such as determining not only the species of plants but also their water and nutrient stress status [36]. This type of applications is only the beginning, as we believe there are many more untapped agriculture and environmental research abilities, such as health-check for crops (disease detection), pest management, irrigation management, crop harvesting, mechanical weeding, and much more. Also, the world population is expected to increase fast, and it is expected to reach about 9 billion by 2050 [14]. Agriculture is in demand for more food. Labor shortages have been reported throughout agriculture in the United States again and again [39]. UGVs can reduce the need for labor in farms by mechanizing and automating agricultural processes such as harvesting, seeding, monitoring, etc. [14].

5.2.1 Fundamental Research Questions

Digital agricultural applications using cognitive swarming of low-cost UGVs require spatial and temporal resolutions. Cognitive swarming of low-cost UGVs is promising but limited by its autonomous navigation controls, management methods, and performances. The fundamental questions are as follows:

- How can smart cognitive algorithms be used to ensure that UGVs can perform state-of-the-art movement in the field?
- How can UGVs/UAVs cooperate efficiently by using a swarming mechanism?
- How can UGVs be taskable with little modification of current planning methodologies?
- How to lower the UGVs barriers (low cost) to entry agricultural industry?
- How to early detect pest population and irrigation non-uniformity mapping based on data information collected from low-cost UGVs?
- How to use data collected from UGV sensors to acquire the most related information about digital agriculture sensing and applications?
- How can UGVs operate safely in emergency situations to avoid hurting people or crops?

5.2.2 *Low Barriers to Entry*

The personal UGV platforms are highly efficient for different agricultural sensing and applications. Specifically, the personal UGV can run more than 6 hours each time and 3 miles per hour, which makes it work more efficiently than most UAVs. Currently, one method to make the UAV fly longer is adding more batteries, such as NCFLY-Helios (Drone Technology Inc.). With eight motors and diameter up to 2136 mm, UAVs can have an endurance of 60 minutes with 1 kilogram payload or 40 minutes with 7 kilogram payload. This is clearly not enough for many agricultural applications in the field. With a single, more cost-effective battery, a personal UGV can work more than 12 hours per day. This makes the UGV platform work much more efficiently in a variety of agricultural proximate sensing and applications. Affordability of the UGV platform is also a very important concern from agricultural industry [2]. Recently, most UGVs used in research are very expensive, such as the IG52-DB4, HD2-S UGV from Superdroid Robots [30], and the Husky UGV from Clearpath Robotics [1]. To make it commercial, balancing technology efficiency with cost is a study area for researchers.

As seen in Table 5.1, compared to Husky UGV, IG52-DB4, and HD2-S UGV, the proposed personal UGV had a much lower cost and better performance. For

Table 5.1 Specifications of several UGVs and the proposed Personal UGV

UGVs	Price	Run time	Max speed	User power
Husky	More than 20,000	3 hours	2.3 mph	5 V, 12 V, and 24 V fused at 5 A each
IG52-DB4	$1613.55	5–5 hours	5.65 mph	24 V, SLA batteries with 18 A capacity
HD2-S	$32,500	up to 8 hours	2.7 mph	24 V, SLA batteries with 18 A of capacity
Personal UGV	Less than $1000	6 hours	3.7 mph	12 V fused at 7 A

example, IG52-DB4 (Superdroid Robots) cost like \$1613.55, and it could only run like 2–5 hours without payload. The proposed personal UGV could run for up to 6 hours. The HD2-S UGV was even more expensive, up to \$32500. The performance was very similar to the proposed UGV platform, which was targeted to less than \$1000. The authors believed that this cost range made the proposed personal UGV achieve lowering barriers to entry and became more ubiquitous in the digital agriculture industry.

As indicated in the above examples, only one payload camera is used to have longer endurance on the UAV platforms. However, for UGV platforms, there is more space for cameras. All kinds of cameras can be mounted on the UGV without being concerned about endurance. For example, the GoPro Hero 6 is the latest product from the GoPro technology company. It can shoot 4K video at 60 frames per second and slow motion 1080P (240 fps), which is only \$399. To make it more cost effective, Gopro Hero 5 can be another choice, which is \$299. For another example, Survey 2 Camera (MAPIR, USA) is a visible light RGB camera. According to its official website, it is only \$280. The Survey 2 has a 16 MP sensor and sharp non-fisheye lens, making it easier to capture in the field. It captures photos on the default timer trigger or can be sent a PWM signal through the HDMI port. Similar size to a GoPro 6 (Gopro, USA), these small survey cameras can be attached to UGV platforms using various mounts. With more cameras, the UGV can become more taskable for farmers and growers.

TurtleBot is a low-cost, personal robot kit with open-source software. It was created at Willow Garage by Melonee Wise and Tully Foote in November 2010. There are mobile base, 3D sensor, and a laptop computer in the TurtleBot kit. TurtleBot has many abilities similar to those of large company robot platforms, such as PR2. With the TurtleBot components, users can even create real-time obstacles avoidance and autonomous navigation. For the personal UGV platform, we can even use the TurtleBot to control the UGVs. The TurtleBot then becomes the "UGV brain" and monitors the UGV behaviors in the field. All the possible applications of the TurtleBot can be realized on the UGVs and can be much more powerful. TurtleBot 2 and TurtleBot 3 are the most commonly used in research. TurtleBot 2 consists of a Kobuki base, a 2200 mAh battery pack, a Kinect sensor, an Asus Xion Pro Live, a fast charger, and a netbook (ROS compatible). ROS stands for Robot Operation System, which is a flexible framework for writing robot software [33]. TurtleBot 3 is a new generation mobile robot which is modular, compact, and customizable. It reduced the size and lowered price without sacrificing function, capability, and quality. The Intel RealSense Depth Camera D435 has been used in several research areas. It can be used indoors and outdoors. The depth technology is active IR stereo. The processor used in this camera is Processor D4. The depth field of view can be up to 1280×720. The depth stream output frame rate can be up to 90 fps. It can be used in 3D point clouds and other research areas. The most important is that the price is only \$179. Mounted on the UGVs, it will be more powerful as part of the cognitive system.

5.2.3 *Cognitive Algorithms by Deep Learning*

Cognitive algorithm is one of the most important parts of the UGV platform. The inspiration of cognitive algorithms for UGV comes from the human drivers. Based on the human cognitive mechanism, researchers lay the foundation for the development of cognitive algorithms. Typically, there are three steps in UGV movement behaviors, stimulus perception, judgment and decision-making, and stimulus performance [10]. In a recent cognitive algorithm, it could identify specific objects in the environment, using cognitive reasoning to develop a deeper understanding of the scene based on objects and their spatial relationships [11]. Reconstruction and mapping monocular simultaneous localization and mapping (SLAM) relies on cognition techniques to calculate the position of the robot and map the environment [16]. Real-time monocular SLAM was thought impossible due to high computational costs. However, there are many types of SLAM packages now, such as LSD-SLAM [8], RatSLAM [25], and SVO [12]. Many open-source vision-based SLAM packages were compared in a paper at 2016 [21]. It turned out that ORB-SLAM is the best performing package [27]. ORB-SLAM uses ORB features, which are binary features that are invariant to rotation and scale [16]. ORB-SLAM also allows robots to be initialized without user input, which is important for UGV platforms as an autonomous robot system. Additionally, when RB-SALM matches features, the number of key frames increases due to the complexity of the images. This is done by "culling" key frames that are no longer used and allows for ORB-SLAM to run for longer periods of time [16].

Deep learning recognition can bridge the gap between perception and intelligence. Perception missions can collect lots of data, but much of it was discarded to simplify interpretation by higher level tasks [38], e.g., a 3D object can become a point in space. In [38], a deep learning framework was used to replace these interfaces with learned interfaces. This allows for error backpropagation that can adapt each module to the robot's task. Deep learning is also a powerful state-of-the-art technique for image processing. It can target land cover and crop type classification from images collected from sensors [20]. Safety is another concern for farmers and growers; the UGV should have the ability to detect workers and farmers and then make the right decision in the field.

5.2.4 *Swarming Mechanism of UGVs*

Technological advances are pushing the applications of UGVs in exciting directions [13]. Many different types of UGVs have been developed to address a variety of agricultural applications ranging from soil sampling, irrigation mapping, and crop harvesting [3]. However, for a large scale of field where fleets of autonomous UGVs must be used to work together on a common goal, multiple UGVs can collaborate to form a swarming mechanism [7].

Recent swarming research has focused on aspects of communications and maintenance of connectivity [37], such as control and collision avoidance with the swarm [42], task allocation and strategies for solving multiple tasks [9], and modeling of the swarm behavior by predicting individual behaviors [41]. In these papers [5, 6, 19, 24], control methodologies for swarms of UGVs were discussed. In this chapter, the research will be conducted based on the proposed UGV platform and provide swarm mechanism with multiple UGVs. In this mode, the followers will follow the leader according to their GPS position. The followers can only follow the leader UGV according to their GPS position, although detailed functions will be developed depending on the requirement of proximate sensing. For example, several UGVs can work in a large-scale field together, which make a mission much shorter than normal time. The collected data can also be shared between UGVs in real-time mode, which makes the UGVs more intelligent.

5.3 Case Study: Build a UGV Platform for Agricultural Research from a Low-Cost Toy Vehicle

5.3.1 Introduction

This section is for a short introduction of the UGV from a toy vehicle (Fig. 5.1). The UGV could navigate through its environment without the assistance or guidance of human intervention. This was done through the signal transmission between the vehicle controller and a program called Mission Planner, which was used to set up markers called waypoints that set up locations for the UGV to reach. For the research purpose, the UGV's main function was to survey and inspect different farmlands, performing large amounts of data collection for researchers to use. After data gathering is complete, they can perform data analysis and computer vision algorithms to gain more insight in crop growth, trends, and pave the way to "digital agriculture."

This tool is a great substitute for physical labor and minimizes expenses in the industry (Fig. 5.2). In addition to research, UGV usage can transition to applications within working sectors such as agriculture, emergency services, government work, and many more. In terms of agriculture, the goal of this project is to create a low-cost (less than $1000) UGV that local farmers in the Central Valley can use. With this affordable UGV, they have a device that does surveillance in their field, inspects for water pipe leakage and crop diseases, and performs human-based tasks such as spraying insecticides and mowing, especially in vineyards. For more information on this UGV application in agriculture, please refer to Chap. 11.

Fig. 5.1 The toy vehicle

Fig. 5.2 The UGV platform

5.4 Chapter Summary

Unlike other robots, UGVs post particular challenges for algorithms of automation, cognition, and swarming. UGVs are operated in an environment with disturbances such as weather conditions, and decisions on control actions have to be made instantaneously. Additionally, farmers and growers are concerned about the profitability of the implementation of UGVs as well as their safety. Therefore, it is necessary to develop a mixed-initiative planning system with adjustable autonomy to allow UGVs to meet their mission objectives within physical dynamics.

More precisely, the architecture of UGV system should include three layers. The first layer includes a team of multiple UGVs with cost-effective cameras and sensors. Each camera sensing a specific band of light is mounted on a single UGV. On this level, it supports direct precise teleoperation for a single UGV by providing guiding mode, loitering mode, or even stabilizing mode controller. In this scenario, the higher layer monitoring system keeps checking the status of the UGV and operation of the user and notifies the user about safety and the corresponding suggestion on action.

The second layer is a model-based monitoring system to improve system safety and operator awareness. The control system is embedded with a declarative model of its activities to detect constraint violations. On the one hand, the performance of UGVs, like position, speed, battery, etc., will be compared with the prediction output based on the prior model of UGV. Once there are any strange behaviors, it will give warning or action suggestions to operators. Particularly, management by consent and management by exception should be involved to free operator's cognitive resources for higher priority operation. On the other hand, during mission planning, human command is also checked to prevent dangerous or conflict initiatives. Not normal interaction would be detected and displayed to the user, and only safe meaningful interaction will pass into the first layer.

The highest level is a mixed-initiative planning system, which allows incremental interactive planning for pest population distribution and non-uniformity irrigation mapping. The scale adaptive searching algorithm is part of this layer for path planning. It orchestrates task assignment, task decomposition, overall path planning, and incremental planning. The user either generates mission plan by task assignment with the planning system or follows the plan sequence with just a few modifications. In addition, it permits incremental planning, so the user can just focus local decisions without endanger the overall system constraints.

References

1. Bhandari, S., Raheja, A., Green, R.L., Do, D.: Towards collaboration between unmanned aerial and ground vehicles for precision agriculture. In: Autonomous Air and Ground Sensing Systems for Agricultural Optimization and Phenotyping II, vol. 10218, p. 1021806. International Society for Optics and Photonics (2017)
2. Billingsley, J., Visala, A., Dunn, M.: Robotics in agriculture and forestry. In: Siciliano, B., Khatib, O. (eds.) Springer Handbook of Robotics. Springer, Berlin (2008). https://doi.org/10.1007/978-3-540-30301-5_47
3. Bonadies, S., Lefcourt, A., Gadsden, S.A.: A survey of unmanned ground vehicles with applications to agricultural and environmental sensing. In: Autonomous Air and Ground Sensing Systems for Agricultural Optimization and Phenotyping, vol. 9866, p. 98660Q. International Society for Optics and Photonics (2016)
4. Burkart, A., Hecht, V., Kraska, T., Rascher, U.: Phenological analysis of unmanned aerial vehicle based time series of barley imagery with high temporal resolution. Precis. Agric. **19**(1), 134–146 (2018)
5. Cai, N., Xi, J.X., Zhong, Y.S.: Swarm stability of high-order linear time-invariant swarm systems. IET Control Theory Appl. **5**(2), 402–408 (2011)
6. Cheah, C.C., Hou, S.P., Slotine, J.J.E.: Region-based shape control for a swarm of robots. Automatica **45**(10), 2406–2411 (2009)

7. Danoy, G., Brust, M.R., Bouvry, P.: Connectivity stability in autonomous multi-level UAV swarms for wide area monitoring. In: Proceedings of the 5th ACM Symposium on Development and Analysis of Intelligent Vehicular Networks and Applications, pp. 1–8 (2015)
8. Engel, J., Schöps, T., Cremers, D.: LSD-SLAM: Large-scale direct monocular SLAM. In: European Conference on Computer Vision, pp. 834–849. Springer, Berlin (2014)
9. Fazli, P., Davoodi, A., Mackworth, A.K.: Multi-robot repeated area coverage. Auton. Robots **34**(4), 251–276 (2013)
10. Feng, L., Gao, L., Li, Y.h.: Research on information processing of intelligent lane-changing behaviors for unmanned ground vehicles. In: 2016 9th International Symposium on Computational Intelligence and Design (ISCID), vol. 2, pp. 38–41. IEEE, Piscataway (2016)
11. Fields, M.A., Lennon, C., Martin, M., Lebiere, C.: Priming for autonomous cognitive systems. In: Micro-and Nanotechnology Sensors, Systems, and Applications IX, vol. 10194, p. 1019421. International Society for Optics and Photonics (2017)
12. Forster, C., Pizzoli, M., Scaramuzza, D.: SVO: fast semi-direct monocular visual odometry. In: 2014 IEEE International Conference on Robotics and Automation (ICRA), pp. 15–22. IEEE, Piscataway (2014)
13. Foster, J.D.: Swarming unmanned aerial vehicles (UAVs): extending marine aviation ground task force communications using UAVs. Tech. rep., Naval Postgraduate School, Monterey, CA, Graduate School of Business and Public (2015)
14. Gebbers, R., Adamchuk, V.I.: Precision agriculture and food security. Science **327**(5967), 828–831 (2010)
15. Ghaffarzadeh, K., Zervos, H.: Conductive ink markets 2013–2018: forecasts, technologies, players. IDTechEx Market Report (2013)
16. Hood, S., Benson, K., Hamod, P., Madison, D., O'Kane, J.M., Rekleitis, I.: Bird's eye view: cooperative exploration by UGV and UAV. In: 2017 International Conference on Unmanned Aircraft Systems (ICUAS), pp. 247–255. IEEE, Piscataway (2017)
17. Kashyap, P.: Machine learning algorithms and their relationship with modern technologies. In: Machine Learning for Decision Makers, pp. 91–136. Springer, Berlin (2017)
18. Kelliher, F.M., Noble, A.D., Dennis, S., Rickerby, A., Knight, T.: A graphical method to evaluate irrigation water application uniformity. N. Z. J. Agric. Res. **60**(1), 80–92 (2017)
19. Kloetzer, M., Belta, C.: Temporal logic planning and control of robotic swarms by hierarchical abstractions. IEEE Trans. Robot. **23**(2), 320–330 (2007)
20. Kussul, N., Lavreniuk, M., Skakun, S., Shelestov, A.: Deep learning classification of land cover and crop types using remote sensing data. IEEE Geosci. Remote Sensing Lett. **14**(5), 778–782 (2017)
21. Li, A.Q., Coskun, A., Doherty, S.M., Ghasemlou, S., Jagtap, A.S., Modasshir, M., Rahman, S., Singh, A., Xanthidis, M., O'Kane, J.M., et al.: Experimental comparison of open source vision-based state estimation algorithms. In: International Symposium on Experimental Robotics, pp. 775–786. Springer, Berlin (2016)
22. Lopes, C.M., Torres, A., Guzman, R., Graça, J., Reyes, M., Vitorino, G., Braga, R., Monteiro, A., Barriguinha, A.: Using an unmanned ground vehicle to scout vineyards for non-intrusive estimation of canopy features and grape yield. In: GiESCO International Meeting, 20th, Sustainable Viticulture and Wine Making in Climate Change Scenarios, 5–10 November 2017. GiESCO (2017)
23. Makori, D., Fombong, A., Abdel-Rahman, E., Nkoba, K., Ongus, J., Irungu, J., Mosomtai, G., Makau, S., Mutanga, O., Odindi, J., et al.: Predicting spatial distribution of key honeybee pests in Kenya using remotely sensed and bioclimatic variables: key honeybee pests distribution models. ISPRS Int. J. Geo-Inform. **6**(3), 66 (2017)
24. Marjovi, A., Marques, L.: Optimal spatial formation of swarm robotic gas sensors in odor plume finding. Auton. Robots **35**(2), 93–109 (2013)
25. Milford, M.J., Wyeth, G.F., Prasser, D.: RatSLAM: a hippocampal model for simultaneous localization and mapping. In: IEEE International Conference on Robotics and Automation, 2004. Proceedings. ICRA'04. 2004, vol. 1, pp. 403–408. IEEE, Piscataway (2004)

26. Muli, E., Patch, H., Frazier, M., Frazier, J., Torto, B., Baumgarten, T., Kilonzo, J., Kimani, J.N., Mumoki, F., Masiga, D., et al.: Evaluation of the distribution and impacts of parasites, pathogens, and pesticides on honey bee (Apis mellifera) populations in East Africa. PLoS One **9**(4), e94459 (2014)
27. Mur-Artal, R., Montiel, J.M.M., Tardos, J.D.: ORB-SLAM: a versatile and accurate monocular SLAM system. IEEE Trans. Robot. **31**(5), 1147–1163 (2015)
28. Oberti, R., Marchi, M., Tirelli, P., Calcante, A., Iriti, M., Tona, E., Hočevar, M., Baur, J., Pfaff, J., Schütz, C., et al.: Selective spraying of grapevines for disease control using a modular agricultural robot. Biosyst. Eng. **146**, 203–215 (2016)
29. Pei, W., Lan, Y., Xiwen, L., Zhiyan, Z., Wang, Z., Wang, Y.: Integrated sensor system for monitoring rice growth conditions based on unmanned ground vehicle system. Int. J. Agric. Biol. Eng. **7**(2), 75 (2014)
30. Pierzchała, M., Giguère, P., Astrup, R.: Mapping forests using an unmanned ground vehicle with 3D LiDAR and graph-SLAM. Comput. Electron. Agric. **145**, 217–225 (2018)
31. Pirk, C.W., Strauss, U., Yusuf, A.A., Demares, F., Human, H.: Honeybee health in Africa—a review. Apidologie **47**(3), 276–300 (2016)
32. Popp, J., Pető, K., Nagy, J.: Pesticide productivity and food security: a review. Agron. Sustain. Dev. **33**(1), 243–255 (2013)
33. Quigley, M., Gerkey, B., Smart, W.D.: Programming Robots with ROS: A Practical Introduction to the Robot Operating System. O'Reilly Media (2015)
34. Rault, R., Trentesaux, D.: Artificial intelligence, autonomous systems and robotics: legal innovations. In: Service Orientation in Holonic and Multi-Agent Manufacturing, pp. 1–9. Springer, Berlin (2018)
35. Romero, D.J., Ben-Dor, E., Demattê, J.A., e Souza, A.B., Vicente, L.E., Tavares, T.R., Martello, M., Strabeli, T.F., da Silva Barros, P.P., Fiorio, P.R., et al.: Internal soil standard method for the Brazilian soil spectral library: performance and proximate analysis. Geoderma **312**, 95–103 (2018)
36. Sankaran, S., Mishra, A., Ehsani, R., Davis, C.: A review of advanced techniques for detecting plant diseases. Comput. Electron. Agric. **72**(1), 1–13 (2010)
37. Saska, M., Baca, T., Thomas, J., Chudoba, J., Preucil, L., Krajnik, T., Faigl, J., Loianno, G., Kumar, V.: System for deployment of groups of unmanned micro aerial vehicles in GPS-denied environments using onboard visual relative localization. Auton. Robots **41**(4), 919–944 (2017)
38. Suppé, A.J., Hebert, M.: Using deep learning to bridge the gap between perception and intelligence. In: Unmanned Systems Technology XIX, vol. 10195, p. 1019503. International Society for Optics and Photonics (2017)
39. Taylor, J.E., Charlton, D., Yúnez-Naude, A.: The end of farm labor abundance. Appl. Econ. Perspect. Policy **34**(4), 587–598 (2012)
40. Tokekar, P., Vander Hook, J., Mulla, D., Isler, V.: Sensor planning for a symbiotic UAV and UGV system for precision agriculture. IEEE Trans. Robot. **32**(6), 1498–1511 (2016)
41. Winfield, A.F., Liu, W., Nembrini, J., Martinoli, A.: Modelling a wireless connected swarm of mobile robots. Swarm Intell. **2**(2), 241–266 (2008)
42. Yu, H., Beard, R.: A vision-based collision avoidance technique for micro air vehicles using local-level frame mapping and path planning. Auton. Robots **34**(1), 93–109 (2013)

Part III
Advanced Big Data Analytics, Plant Physiology-Informed Machine Learning, and Fractional-Order Thinking

Chapter 6
Fundamentals of Big Data, Machine Learning, and Computer Vision Workflow

Abstract This chapter serves as a foundational guide to the essential principles of big data, machine learning, and computer vision workflows. The exploration begins with an introduction, setting the stage for a comprehensive tutorial that elucidates the workflow's fundamentals. The chapter unfolds with a step-by-step tutorial focused on the classification of cotton water stress using Convolutional Neural Networks (CNNs).

6.1 Introduction

In today's data-centric world, the confluence of Big Data, Machine Learning, and Computer Vision has ignited a revolution in our ability to make sense of vast and complex datasets. This chapter serves as a foundational exploration into the intricate interplay of these three dynamic fields, setting the stage for a comprehensive understanding of the workflow that underpins them. Big Data, the cornerstone of this triad, introduces the concept of managing and analyzing enormous volumes of data, uncovering valuable insights and patterns. Machine learning, the driving force behind intelligent decision-making, equips machines with the ability to learn, adapt, and predict based on this wealth of information. Finally, Computer Vision gives machines the ability to interpret visual data, bridging the gap between the digital world and the physical world. The primary emphasis of this chapter is placed on the pragmatic application of machine learning and computer vision with a deliberate focus on a real-world scenario, cotton water stress classification, rather than extensively delving into the intricate theoretical underpinnings of these domains. Our focus is on elucidating the six critical steps that comprise the cotton water stress classification workflow, including Data Loading, Data Preprocessing , Train and Test Split, Normalization, Training the Model, and Model Evaluation. These steps serve as a pivotal guidepost for the application of ML and CV techniques, illuminating their practical significance and relevance in the real world.

H. Niu, Y. Chen, *Smart Big Data in Digital Agriculture Applications*, Agriculture Automation and Control, https://doi.org/10.1007/978-3-031-52645-9_6

1. Data Loading
   The authors will begin the journey by addressing the fundamental step of Data Loading. This step involves gathering and acquiring the data necessary for one's ML or CV project. The authors will discuss data sources, formats, and the importance of raw data quality.
2. Data Preprocessing
   Moving forward, the authors will delve into Data Preprocessing. Here, the authors will explore techniques for cleaning and preparing the data to ensure it is in the best possible condition for analysis.
3. Train and Test Split
   The next critical step is Train and Test Split. The authors explain the importance of dividing their dataset into separate training and testing subsets, ensuring that the model's performance can be accurately assessed.
4. Normalization
   Normalization is another pivotal concept that will be explored. The authors discuss how scaling the data can enhance a model performance and why it is essential in various ML and CV applications.
5. Training the Model
   With data prepared, the authors dive into the heart of the matter, training the model. One will gain insight into the practical aspects of training ML and CV models, including model selection and hyperparameter tuning.
6. Model Performance Evaluation
   Finally, the journey culminates in evaluating Model Performance. The authors explore the metrics and techniques used to assess the effectiveness of the model and discuss strategies for improving the accuracy and generalizability.

Before moving on, let us review the research background being used for this chapter: Research focused on irrigation management is gaining momentum, as it aims to achieve sustainability and enhance marketability in the field of agriculture, which is responsible for more than 70% of global water consumption. Efficient water usage through irrigation has always played a pivotal role in the advancement of agriculture and successful farming practices. Embracing smart irrigation management techniques empowers growers to irrigate with greater efficiency, thus promoting sustainable agricultural production. In this context, growers often rely on crop evapotranspiration (ET_c) as a key factor in making informed irrigation decisions, underscoring the significance of accurately determining and spatially mapping crop water status [1]. Technological progress, exemplified by the emergence of Unmanned Aerial Vehicles (UAVs), has brought about a revolutionary shift in agricultural monitoring. UAV platforms are capable of capturing high-resolution images with centimeter-level spatial accuracy and offer higher temporal coverage compared to satellite imagery [4, 7]. In light of these advances, this study introduces a robust method for classifying water stress in cotton using a compact UAV platform and convolutional neural networks (CNNs). The experiment was conducted at the USDA-ARS Cropping Systems Research Laboratory (CSRL) in Lubbock, Texas, where the cotton field was divided into 12 drip zones (Fig. 6.1). The study

Fig. 6.1 The cotton field experimental layout. The experiment was conducted at the USDA-ARS Cropping Systems Research Laboratory (CSRL) in Lubbock, Texas, where the cotton field was divided into 12 drip zones. The study included three replications to evaluate four irrigation treatments: "rainfed," "full irrigation," "percent deficit of full irrigation," and "time delay of full irrigation"

included three replications to evaluate four irrigation treatments: "rainfed," "full irrigation," "percent deficit of full irrigation," and "time delay of full irrigation." The results demonstrated that the CNN model successfully classified cotton water stress using the UAV-based RGB image, achieving an overall best prediction accuracy of approximately 91%. The research findings highlighted the state-of-the-art performance of the proposed system in the classification of cotton water stress and provided valuable information on key image characteristics that contribute to the accurate classification. The authors concluded that the integration of UAV-based RGB imagery and CNN models had great potential to assess water stress in cotton.

Throughout this chapter, our aim is to equip one with a practical understanding of the six essential steps mentioned above, providing one with the tools and knowledge to apply machine learning and computer vision effectively in their own projects. Whether you are a novice or an experienced practitioner, the authors' focus is on the practical, real-world application, such as the above cotton water stress classification task, of these techniques to solve problems and make informed decisions.

6.2 A Fundamental Tutorial: Cotton Water Stress Classification with CNN

6.2.1 Data Loading

The initial step in machine learning and computer vision workflow is Dataset Loading, where data are collected and acquired to serve as the foundation of the entire analysis. This phase is often underestimated, but its significance cannot be overstated, as the quality of the data fundamentally impacts the validity and reliability of any subsequent analysis or model. Ensuring data integrity, relevance, and quality is crucial to laying a strong foundation for successful machine learning and computer vision endeavors. Without the data, the entire analytical process may be compromised, emphasizing the pivotal role of Dataset Loading in the overall workflow.

Here in the authors' example, the dataset is loaded from the Google Drive.[1] First of all, the authors import the essential packages needed for this workflow.

```
import numpy as np
import pandas as pd
import sklearn
from sklearn.model_selection import train_test_split
import matplotlib.pylab as plt
import tensorflow as tf
from tensorflow.keras import datasets, layers,
    models
import sys
import glob
import skimage.color
import skimage.filters
from skimage.morphology import opening, closing,
    erosion, dilation
from skimage.morphology import disk
from skimage.color import rgb2lab

from google.colab import drive
drive.mount('/content/drive/MyDrive/Colab Notebooks'
    )
```

The following code block shows the dataset was saved as a .npy file in the Google Drive. There are 6832 cotton RGB images with a size of 64 × 64 × 3. The high-resolution UAV images were collected from September 2, 2022. The corresponding label .csv file was also loaded with class names.

```
data = np.load('/content/drive/MyDrive/Colab
    Notebooks/twri_rgb_6832_cotton_64x64_9_02.npy')
labels = pd.read_csv('labels_rgb.csv')
```

[1] The code is available at https://github.com/hniu-tamu/cotton_water_stress_remote_sensing.

```
class_names = ['rainfed', 'fully irrigated', 'percent deficit', 'time delay']
```

6.2.2 Data Preprocessing

Data Preprocessing is a critical phase in the machine learning and computer vision workflow, playing a foundational role in ensuring that data are primed for analysis. This phase involves a series of operations that transform and refine the raw dataset, making it more suitable for the specific ML or CV task at hand. Data Preprocessing encompasses various essential procedures, including handling missing values, addressing outliers, and standardizing data formats. Additionally, it involves the transformation of categorical data into numerical representations, ensuring that the data are in a format that ML and CV models can effectively work with. By addressing these intricacies, Data Preprocessing sets the stage for more accurate and reliable model training and analysis, paving the way for successful outcomes in ML and CV applications.

In this example, the authors show how to remove unwanted dataset. In the original loaded dataset, there is one class called "filler," which was the buffer areas in the cotton field. The authors would like to remove it from the dataset. Inside the for loop, there is an if statement: if val != 0:. This condition checks whether the label at the current index is not equal to 0. In this context, a label of 0 indicates a "filler" or unwanted data point. The y_label is also modified by subtracting 1 from each element in the "new_y_label_without_filler" array. This is also a preprocessing step to adjust the label values because the authors want the new labels to be 0-based.

```
new_data_without_filler = np.zeros((5376, 64, 64, 3))
new_y_label_without_filler = np.zeros(5376, dtype=int)
i = 0
for idx, val in enumerate(y_label):
  if val != 0:
    new_data_without_filler[i] = data[idx]
    new_y_label_without_filler[i] = y_label[idx]
    i+=1
data = new_data_without_filler
y_label = new_y_label_without_filler - 1
```

After the preprocessing steps have been applied to the cotton image dataset, it is essential to assess the quality of the data and ensure that it is ready to use in machine learning or data analysis tasks. One effective way to do this is through visualization. Visualizing the data provides a human-friendly way to confirm that the preprocessing steps were successful and to gain insights into the dataset's characteristics. Given that the cotton image dataset is likely large, it is often practical to visualize a representative subset of the images rather than the entire dataset. For instance, one might choose to display 25 images to give a sense of the data quality and diversity. To visualize the selected images, one can make use of various data visualization libraries in a preferred programming language. Python, for example, offers a range of libraries like Matplotlib or Seaborn. These libraries allow one to display images and customize the presentation. Here, the authors arrange the selected images in a grid format, a 5 × 5 grid in the case of displaying 25 images. This format is visually appealing and provides a clear structure for viewing the images (Fig. 6.2).

```
plt.figure(figsize=(10,10))
for i in range(25):
    plt.subplot(5,5,i+1)
    plt.xticks([])
    plt.yticks([])
    plt.grid(False)
    plt.imshow(data[i]/255)
    plt.xlabel(class_names[y_label[i]])
plt.show()
```

To further explore the influence of specific factors on the prediction of water stress, such as cotton canopy and soil characteristics, the authors turned to morphological image processing methods using scikit-image, an image processing package in Python.[2] Morphological image processing is a fundamental technique for analyzing and manipulating images based on their shape and structure [2]. It is rooted in mathematical morphology, which provides a set of operations to extract, enhance, and modify the geometric features present in an image. These operations are primarily based on the concepts of dilation, erosion, opening, and closing [6]. Dilation expands the shape boundaries and fills gaps, while erosion erodes the boundaries and removes small details. The opening is a combination of erosion followed by dilation, which helps remove noise and small objects. In contrast, closing is a combination of dilation followed by erosion, useful for filling gaps, and connecting broken structures [2].

Using morphological image processing techniques, it becomes possible to extract important features, remove noise, segment objects, and perform other essential image analysis tasks, ultimately aiding in the interpretation and understanding of the cotton image (Fig. 6.3). By segmenting the image into cotton and soil areas, the authors could focus on studying the distinct properties and characteristics of each region. This approach enables a comprehensive analysis of how water stress manifests within the cotton canopy versus the soil, providing valuable information on the factors that influence the availability and distribution of water within the field.

To accomplish the task of segmenting the cotton canopy, the authors initially converted the RGB image into the LAB color space. The LAB color space is a color model that represents colors based on three components: L (lightness), A (green–red), and B (blue–yellow). By transforming the RGB image into LAB color space, the authors aimed to exploit the distinctive characteristics of the LAB color channels for better differentiation between the cotton canopy and the background. Due to the

[2] https://scikit-image.org/.

Fig. 6.2 A 5 × 5 grid in the case of displaying 25 cotton images

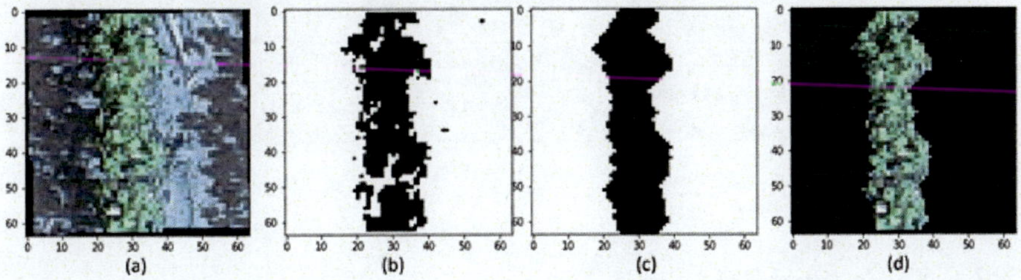

Fig. 6.3 Cotton segmentation with morphological image processing methods. **(a)** The original UAV-based RGB image, **(b)** a rough mask image after Otsu's methods, **(c)** the mask after opening and dilation operations, and **(d)** the cotton canopy image with the mask in **(c)**

green color of the cotton canopy, the authors mainly chose the A channel as the grayscale image for further image processing.

To obtain a rough mask image for cotton canopy segmentation, the authors employed Otsu's method, a widely used technique for automatic thresholding [5].

Otsu's method calculates an optimal threshold value by maximizing the between-class variance of a grayscale image. The resulting rough mask image provided an initial segmentation of the cotton canopy region, albeit with some imperfections and noise (Fig. 6.3b). This rough mask served as a starting point for further refinement and fine-tuning of the cotton canopy segmentation process. Subsequent morphological operations, such as opening and dilation, could be applied to improve the accuracy and smoothness of the final mask image (Fig. 6.3c and d). The conversion of the RGB image into LAB color space and subsequent segmentation of the cotton canopy using morphological image processing techniques provided the authors with a targeted representation of the cotton canopy region. After applying Otsu's method, opening, and dilation operations to generate mask images, the authors manually assessed the segmentation performance (Fig. 6.4). This segmentation step was essential for further analysis and accurate prediction of water stress in the cotton field, as it isolated the canopy region from the surrounding background and facilitated the extraction of relevant characteristics for subsequent classification tasks. This part of the code is shown below:

```
# Morphology image processing, keep cotton only for i in range(len (data)): # Convert rgb to lab color space
    image = data[i]/255.0
    gray_image = rgb2lab(image)
    gray_image = gray_image[:,:,1]

    # Use Otsu's method to get a rough mask image
    val = skimage.filters.threshold_otsu(gray_image)
    mask = gray_image > val

    footprint = disk(2)
    mask = opening(mask, footprint)
    mask = dilation(mask, disk(1))

    selection = image.copy()
    selection[mask] = 0
    data[i] = selection
```

6.2.3 Train and Test Split

The Train and Test Split is a pivotal phase in the machine learning workflow, instrumental in the assessment of a model's performance. It involves the division of the dataset into two distinct subsets: the training set and the testing set. The significance of this step lies in its ability to ensure that the model performance is evaluated accurately and rigorously. The training set serves as the foundation upon which the model is built, allowing it to learn and adapt to the underlying patterns within the data. In contrast, the testing set serves as an independent dataset that

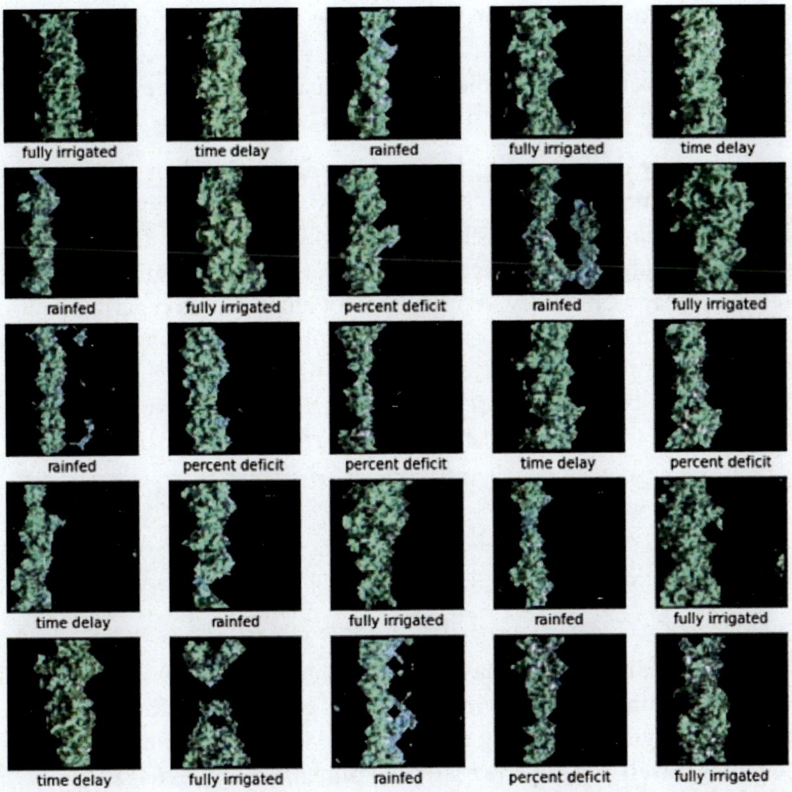

Fig. 6.4 A visual demonstration of an image dataset after morphological image processing

the model has never encountered before, providing a reliable gauge of the model's predictive power. By assessing the model performance on unseen data, the Train and Test Split step ensures that the model can generalize well to new, real-world situations. It aids in detecting potential issues such as overfitting, where the model performs well on the training data but poorly on new data. Therefore, the Train and Test Split is an indispensable safeguard in the model evaluation process, guiding the development of robust and reliable machine learning solutions.

When working with images for deep learning, one needs both the images themselves, usually denoted X, and also correct labels for these images, usually denoted Y. Furthermore, one needs X and Y values both for training the model and then a separate set of X and Y values for validating the performance of the model after it has been trained. Therefore, four segments of data are needed for the cotton image dataset:

1. x_train: Images used for training the CNN model.
2. y_train: Correct labels for the x_train images, used to evaluate the model's predictions during training.

3. *x_valid*: Images set aside for validating the performance of the model after it has been trained.
4. *y_valid*: Correct labels for the *x_valid* images used to evaluate the model's predictions after it has been trained.

There are many deep learning frameworks, each with their own merits. In this chapter, one will be working with TensorFlow 2 and specifically with the Keras API. Keras has many useful built-in functions designed for the computer vision tasks. It is also a legitimate choice for deep learning in a professional setting due to its readability and efficiency, though it is not alone in this regard, and it is worth investigating a variety of frameworks when beginning a deep learning project. One of the many useful features that Keras provides is modules containing many helper methods for many common datasets [3]. One will begin by partitioning the cotton image dataset into training and testing with the *train_test_split* method.

```
# The data, split between train and validation sets
train_images, test_images, train_labels, test_labels
    = train_test_split(data, y_label, test_size
    =0.3, random_state=42)
train_images, test_images = train_images / 255.0,
    test_images / 255.0
```

As stated above, the cotton image dataset contained 5376 RGB images of cotton canopy. By executing the following cells, one can see that Keras has partitioned 3763 of these images for training and 1613 for validation (after training) and also that each image itself is a 3D array with the dimensions $64 \times 64 \times 3$:

```
train_images.shape
#(3763, 64, 64, 3)
test_images.shape
#(1613, 64, 64, 3)
```

Furthermore, one can see that these $64 \times 64 \times 3$ images are represented as a collection of float64 values between 0.0 and 1.0.

```
train_images.dtype
# dtype('float64')
test_images.shape
# 0.0
train_images.max()
# 1.0
```

6.2.4 Creating the Model

With the data prepared for training, it is now the time to create the CNN model that we will train with the data. This first basic model will be made up of several layers and will be comprised of three main parts:

1. An input layer, which will receive data in some expected format.

2. Several hidden layers, each comprised of many neurons. Each neuron will have the ability to affect the network's estimate with its weights, which are values that will be updated over many iterations as the network gets feedback on its performance and learns.
3. An output layer, which will depict the network's estimate for a given image.

To begin, one will use Keras *Sequential* model class to instantiate an instance of a model that will have a series of layers that data will pass through in sequence:

```
model = models.Sequential()
```

Next, let us add the input *Conv2D* layers and *MaxPooling2D* layers. The *Conv2D* layers are specifically designed to process two-dimensional data, such as images or grids, and they play a crucial role in capturing patterns and features from the data. Here is an explanation of a *Conv2D* layer.

1. Convolution operation
   The primary purpose of a *Conv2D* layer is to perform convolution operations on the input data. Convolution is a mathematical operation that involves sliding a small filter (also known as a kernel) over the input data to extract local patterns and features. This operation helps the network detect edges, textures, and other visual characteristics.
2. Filters and feature maps
   A *Conv2D* layer uses a set of learnable filters, which are small 2D matrices. These filters are initialized with random values and are updated during training through backpropagation. Each filter is responsible for detecting a specific type of feature in the input data.
3. Stride
   The stride is a parameter that defines how the filter slides over the input data. It determines the step size of the filter as it moves across the input. A larger stride results in a smaller output feature map with reduced spatial dimensions, while a smaller stride leads to a larger output feature map.
4. Padding
   Padding is another parameter that can be applied to control the spatial dimensions of the output feature map. Two common types of padding are "valid" and "same." "Valid" padding means no padding is added, and the output feature map will be smaller than the input. "Same" padding adds zero values around the input, ensuring that the output has the same spatial dimensions as the input.
5. Activation function
   After the convolution operation, each element of the output feature map is passed through an activation function, typically a Rectified Linear Unit (ReLU) function. This introduces nonlinearity into the network and helps it capture complex patterns.
6. Multiple filters
   Conv2D layers usually have multiple filters (also called channels), allowing the network to capture a variety of features simultaneously. The output feature map will have as many channels as there are filters in the *Conv2D* layer.

7. Learning
   During training, the neural network adjusts the filter weights to minimize the error between the predicted output and the actual target values. This is achieved through backpropagation and gradient descent.

Utilizing the *relu* activation function by the authors serves to enhance the network's ability to generate more nuanced predictions about the data, surpassing the limitations imposed by a strictly linear function. The *input_shape* parameter defines the format of the incoming data, here denoted as a 3D array with dimensions of 64 × 64 × 3 in this specific context:

```
model.add(layers.Conv2D(32, (3, 3), activation='relu
    ', input_shape=(64, 64, 3)))
model.add(layers.MaxPooling2D((2, 2)))
model.add(layers.Conv2D(64, (3, 3), activation='relu
    '))
model.add(layers.MaxPooling2D((2, 2)))
model.add(layers.Conv2D(64, (3, 3), activation='relu
    '))
```

The *MaxPooling2D* is a downsampling operation commonly used in CNN to reduce the spatial dimensions of feature maps while retaining important information. It is typically applied after convolutional layers to reduce the computational load and improve the network's ability to capture essential features. Then, the authors add an additional densely connected layer that gives the network more parameters to contribute toward its guesses and therefore more subtle opportunities for accurate learning:

```
model.add(layers.Flatten())
model.add(layers.Dense(64, activation='relu'))
```

Finally, the authors add an output layer. This layer results in each of the layer's values being a categorical probability between 0 and 1 and will result in all the outputs of the layer adding to 1. In this case, since the network is to make a guess about a single image belonging to one of four possible irrigation levels, there will be four outputs. Each output gives the model's guess (a probability) that the image belongs to that specific class:

```
model.add(layers.Dense(4))
```

If interested in plotting the model summary (Fig. 6.5), one can use the following code for that:

```
from keras.utils import plot_model
plot_model(model, to_file = 'model_sammary.pdf',
    show_shapes=True, show_layer_names= True)
```

The final step that the authors need to take before one can actually train the CNN model with data is to compile it. Here the authors specify a loss function that will be used for the model to understand how well it is performed during training. It also specifies that the authors would like to track accuracy while the model trains.

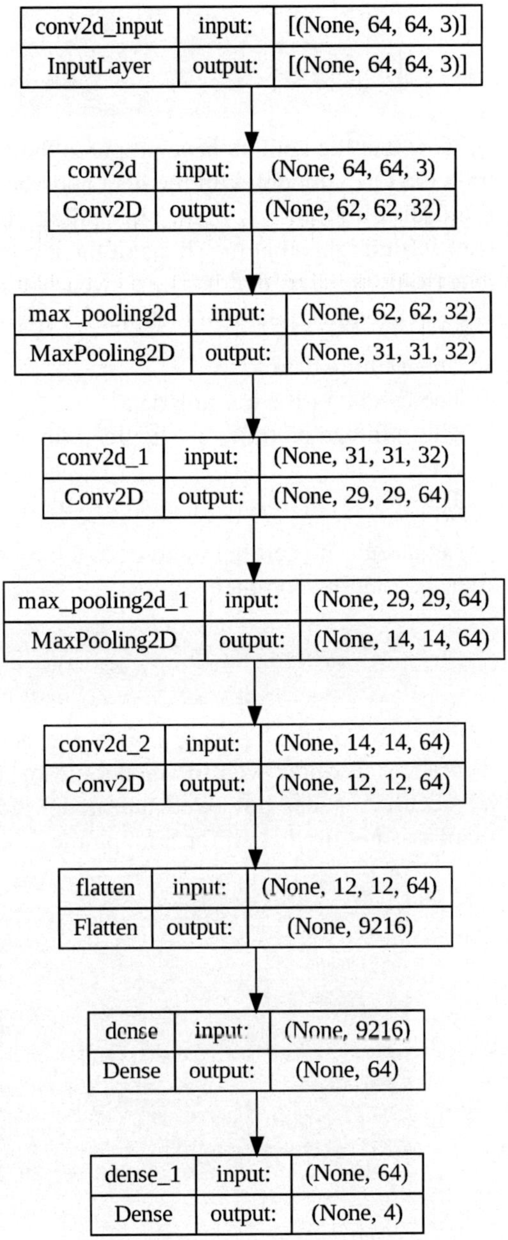

Fig. 6.5 The CNN model summary. This architecture represents a CNN designed for a specific task, where the input images are of size 64 × 64 pixels with three color channels. The network includes convolutional layers with max pooling to extract hierarchical features, followed by fully connected (dense) layers for classification

```
model.compile(optimizer='adam',
              loss=tf.keras.losses.
  SparseCategoricalCrossentropy(from_logits=True),
              metrics=['accuracy'])
```

Now that the authors have prepared the training and validation data, it is time to train the CNN model with the data and verify it with its validation data. "Training a model with data" is often also called "fitting a model to data." Put this latter way, it highlights that the shape of the model changes over time to more accurately understand the data that is being given. When fitting (training) a model with Keras, one can use the model's fit method. It expects the following arguments:

1. The training data
2. The labels for the training data
3. The number of times it should train on the entire training dataset (called an epoch)
4. The validation or test data and its labels

Let us run the cell below to train the model. The authors will discuss its output after the training completes:

```
history = model.fit(train_images, train_labels,
    epochs=70, validation_data=(test_images,
    test_labels))
```

For each of the epochs, notice the accuracy and val_accuracy scores. The accuracy states how well the model did for the epoch on all the training data. The val_accuracy states how well the model did on the validation data, which if you recall was not used at all for training the model (Fig. 6.6).

```
plt.plot(history.history['accuracy'], label='
    accuracy')
plt.plot(history.history['val_accuracy'], label = '
    val_accuracy')
plt.xlabel('Epoch')
plt.ylabel('Accuracy')
plt.ylim([0.5, 1.2])
plt.legend(loc='lower right')

test_loss, test_acc = model.evaluate(test_images,
    test_labels, verbose=2)
```

6.2.5 *The Model Performance Evaluation*

To evaluate the performance of trained CNN models, one can employ a confusion matrix (Fig. 6.7), which provides a comprehensive summary of the prediction results in a classification problem. The confusion matrix helps to understand not only the overall accuracy of the classifier but also the specific types of errors that are made.

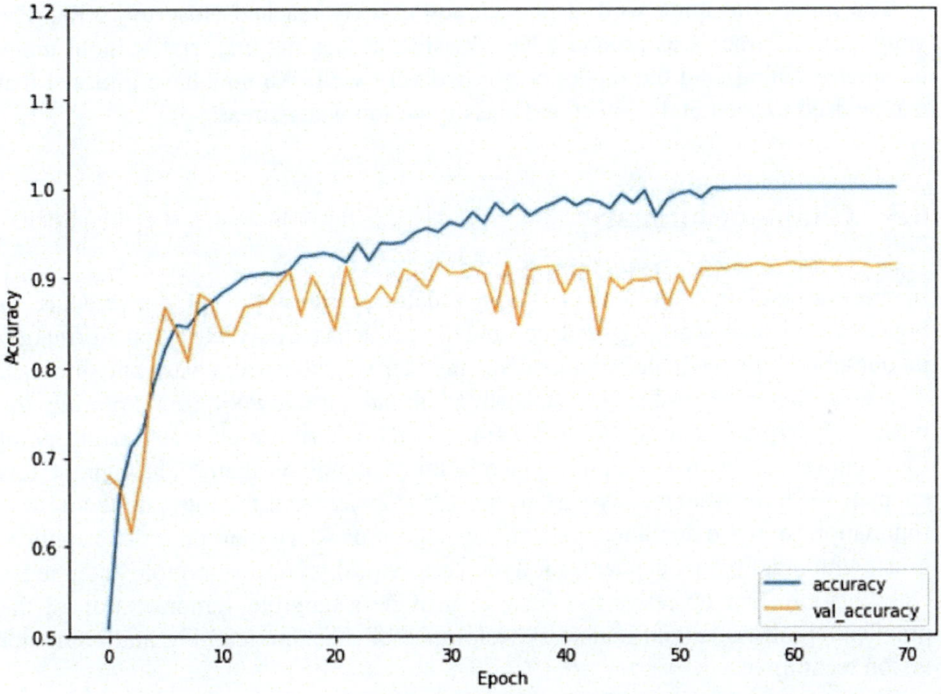

Fig. 6.6 The training and testing accuracy of the CNN model

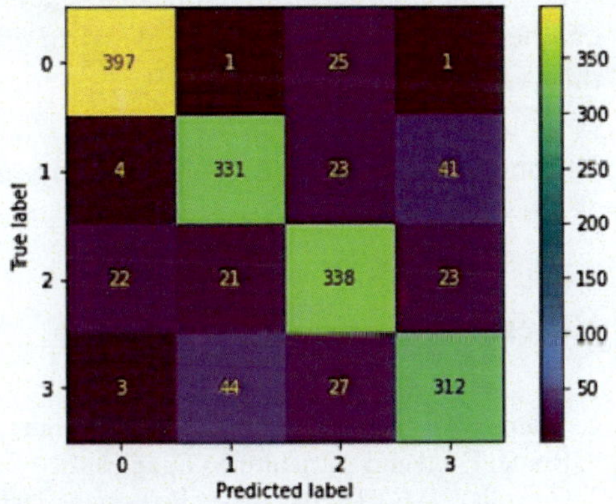

Fig. 6.7 The summary of prediction results of the CNN model for original image dataset at different sampling days. The irrigation treatments were labeled as follows: "rainfed" as "0," "fully irrigated" as "1," "percent deficit" as "2," and "time delay" as "3." The color bar on the right side represented the number of test samples

```
probability_model = tf.keras.Sequential([model, tf.
    keras.layers.Softmax()])
predictions = probability_model.predict(test_images)
pred = np.argmax(predictions, axis=1)
tf.math.confusion_matrix(test_labels, pred)

from sklearn.metrics import ConfusionMatrixDisplay
ConfusionMatrixDisplay.from_predictions(test_labels,
     pred)
```

The model did quite well. The accuracy quickly reached close to 100%. The validation accuracy was around 87%. Consider it was the first trial without many parameter tuning, and the model performed very well. We now have a model that can be used to accurately detect and classify cotton water stress.

6.3 Chapter Summary

In the ever-evolving landscape of today's data-centric world, the convergence of big data, machine learning, and computer vision has catalyzed a transformation in our ability to navigate, understand, and derive value from vast and intricate datasets. This chapter has laid the foundational groundwork for exploring the intricate interplay of these three dynamic fields, offering a critical starting point for comprehending the underlying workflow that unifies them. Throughout this chapter, we have strategically shifted our focus away from the intricate theoretical foundations of these domains. Instead, we have emphasized their pragmatic applications, with a deliberate concentration on a real-world scenario—cotton water stress classification. Our objective has been to provide a tangible demonstration of the practical significance and real-world relevance of machine learning and computer vision techniques.

The spotlight has been on elucidating the six critical steps that constitute the cotton water stress classification workflow. These steps not only serve as a pivotal guidepost for practitioners but also exemplify how these cutting-edge technologies can be harnessed to address real-world challenges. By understanding and mastering this workflow, we can harness the power of big data, machine learning, and computer vision to make informed, data-driven decisions and contribute to the ongoing data revolution that defines our era.

References

1. Allen, R.G., Pereira, L.S., Raes, D., Smith, M.: FAO Irrigation and Drainage Paper No. 56. Rome: Food and Agriculture Organization of the United Nations **56**(97), e156 (1998)
2. Gonzalez, R.C.: Digital Image Processing Using MATLAB, 3rd edn. Gatesmark (2020)
3. Manaswi, N.K., Manaswi, N.K.: Understanding and working with Keras. Deep Learning with Applications Using Python: Chatbots and Face, Object, and Speech Recognition with TensorFlow and Keras, pp. 31–43 (2018)
4. Niu, H., Zhao, T., Wang, D., Chen, Y.: A UAV resolution and waveband aware path planning for onion irrigation treatments inference. In: 2019 International Conference on Unmanned Aircraft Systems (ICUAS), pp. 808–812. IEEE, Piscataway (2019)
5. Otsu, N.: A threshold selection method from gray-level histograms. IEEE Trans. Syst. Man Cybern. **9**(1), 62–66 (1979)
6. Szeliski, R.: Computer Vision: Algorithms and Applications. Springer, Berlin (2022)
7. Zhao, T., Koumis, A., Niu, H., Wang, D., Chen, Y.: Onion irrigation treatment inference using a low-cost hyperspectral scanner. In: Multispectral, Hyperspectral, and Ultraspectral Remote Sensing Technology, Techniques and Applications VII. International Society for Optics and Photonics (2018)

Chapter 7
A Low-Cost Proximate Sensing Method for Early Detection of Nematodes in Walnut Using Machine Learning Algorithms

Abstract This chapter presents an innovative low-cost proximate sensing method designed for the early detection of nematodes in walnut trees, leveraging machine learning algorithms. The chapter commences with an introduction, highlighting the significance of early detection in managing nematode infestations and introducing the approach's cost-effective nature. The Materials and Methods section outlines the study area, emphasizing the use of reflectance measurements with a radio frequency sensor for data collection. The ground truth data collection and processing methodologies are detailed, providing a transparent overview of the experimental setup. The chapter introduces classification algorithms from scikit-learn, along with the application of deep neural networks (DNNs) using TensorFlow. The Results and discussion section follows, including data visualizations for Project 45 in 2019, the performance of classifiers in both Project 45 instances in 2019 and 2020. The findings offer insights into the effectiveness of the low-cost proximate sensing method for early nematode detection, providing a foundation for further exploration and refinement. The chapter concludes with a comprehensive summary, encapsulating key findings and implications of the low-cost sensing approach.

7.1 Introduction

Most of the high-value commodity walnuts of the USA are grown in the fertile soils of the California Central Valley. The sustainability of walnut production is challenged by several pests and pathogen problems that can occur in walnut orchards [3, 20]. The plant-parasitic nematodes are one of these pest problems. There are many species of nematodes in California, but the root-knot nematodes (*Meloidogyne* spp.) and root-lesion nematodes (*Pratylenchus vulnus*) [35] are among the most damaging [34]. Plant-parasitic nematodes are difficult to control and can spread easily in soil on tools, boots, or infested plants [30]. Root-knot nematodes can attack many different crops, including nut and fruit trees. Root-knot nematodes usually cause distinctive swellings, called galls, on the roots of affected plants [5]. Many nut crop rootstocks used in California carry resistance toward root-knot nematodes,

H. Niu, Y. Chen, *Smart Big Data in Digital Agriculture Applications*, Agriculture Automation and Control, https://doi.org/10.1007/978-3-031-52645-9_7

but nematode species new to California can occur in the state and damage the nut crops based on their vulnerability. Root-lesion nematodes are also widely distributed and heavily damaging in walnut orchards, therefore causing widespread problems [19].

Recently, accurate nematode detection relies on soil collections, tedious nematode extraction procedures, and identification and enumeration under a microscope [35]. Infestations of nematode under field conditions are not diagnostic but need to be confirmed by laboratory procedures. These traditional nematode detection methods are tedious, need highly specialized personnel, equipment, and take time to complete. The detection of plant-parasitic nematodes by simple means is paramount to direct management strategy decisions.

With the development of remote sensing technology, satellite and unmanned aerial vehicles (UAVs) imagery have been commonly used in digital agriculture, such as early detection of nematodes [15, 18, 26], irrigation management [24, 37], and evapotranspiration estimation [21–23]. The walnut root infection by nematodes can induce spectral variation in leaves and define a special spatial configuration. Thus, Hillnhütter et al. reported that the induction of symptoms in leaves, the clustered occurrence, and the low level of mobility in the soil made nematodes ideal targets for remote sensing detection [10]. For example, Martins et al. detected and mapped the root-knot nematode infection in coffee crops using the hyperspectral data and RapidEye sensor to identify the most sensitive spectral ranges for nematode infection discrimination in coffee plants [18]. The multispectral classification method could classify the spatial distribution of healthy, moderately infected, and severely infected coffee plants with an overall accuracy of 78% [18].

However, satellite or UAVs could be limited by flight time or payload capability. Such limitations do not apply to proximate sensors. For instance, in [16], Li et al. adopted a non-invasive method of measuring leaf water content using terahertz (THz) radiation, which provided for repeated, nondestructive measurement of leaf water content. In [25], the authors proposed a novel low-cost proximate radio frequency tridimensional sensor Walabot to measure the stem water potential of walnut leaves. Results showed that the sensor could predict stem water potentials with an accuracy of 78% using the decision tree classifier. The Walabot has also been used in many other research topics, such as battery management [32, 33]. For example, Wang et al. [33] compared the performance of three classifiers on Lithium-ion batteries (LIBs) voltage classification problem, which were principal component analysis (PCA) [12], linear discriminant analysis (LDA) [6], and stochastic gradient descent (SGD) classifiers. Experimental results showed that LDA was the most suitable for LIBs voltage classification. These findings highlighted the powerful potential for applying Walabot as a proximate sensor.

Inspired by previous research work, the **objectives of this chapter** are as follows:

1. Investigate and validate the non-invasive approach of early detection of nematodes using a pocket-sized, cutting-edge technology radio frequency tridimensional sensor.
2. Implement several scikit-learn classifiers, such as "Neural Net" [28], "Support Vector Machine" [4], "Random Forest" [2], "AdaBoost" [14], "Nearest Neighbors" [8], and "Decision Tree" [17] to classify the nematode infection levels of walnut trees.
3. Evaluate the prediction performance of the Deep Neural Networks (DNNs) model.

The **major contributions of this chapter** are as follows:

1. Develop a reliable method for early detection of nematodes using a proximate sensor and ML algorithms.
2. Provide a DNN framework, which establish a DNN model between the radio frequency reflectance of walnut leaves and nematode infection levels.

The rest of this chapter is organized as follows: Section 7.2 introduces materials and methods for early detection of nematode infection levels. Several commonly used ML algorithms are also introduced in this section. Different results of early detection of nematode infection levels are then presented, compared, and discussed in Sect. 7.3. In the end, the authors share views on the early detection of nematodes with ML algorithms for future research and draw conclusive remarks.

7.2 Materials and Methods

7.2.1 Study Area

The study was conducted in an experimental walnut orchard at the Kearney Agricultural Research and Extension Center, Parlier, CA, USA (36.595°N, 119.508°W). There were three replications to detect the nematode infection levels, Project 30 (a) and (b), Project 45, and Project Smith (Fig. 7.1). For example, in Project 45, the experimental orchard consisted of 90 trees in total, including experimental clonal hybrid walnut rootstock genotypes (originating from one cross) and some clonal controls (Fig. 7.1b). They foremost varied for their host status toward root-lesion nematode.

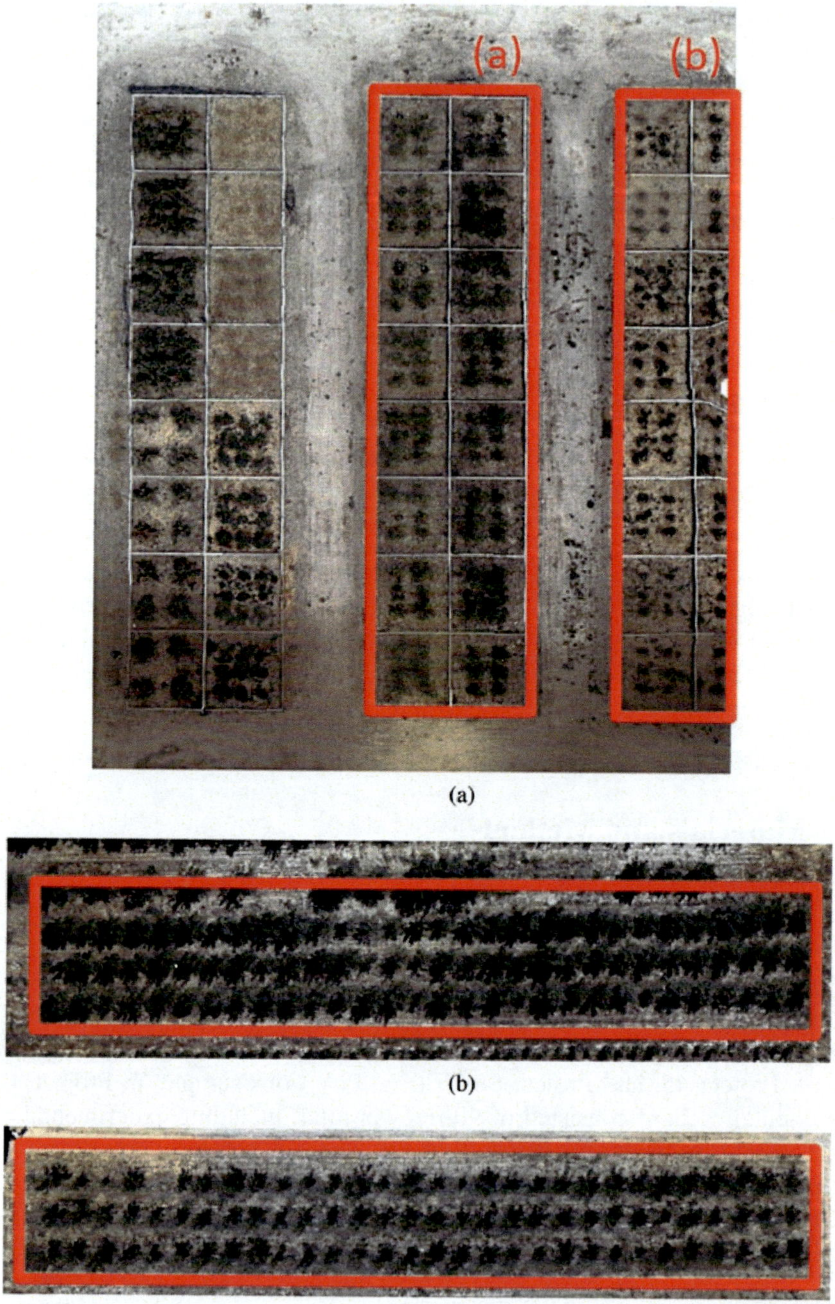

(a)

(b)

(c)

Fig. 7.1 The study was conducted in an experimental walnut orchard at the Kearney Agricultural Research and Extension Center, Parlier, CA, USA (36.595°N, 119.508°W). There were three replications to detect the nematode infection levels. (**a**) Project 30 (a, b), (**b**) Project 45, and (**c**) Project Smith

7.2.2 Reflectance Measurements with a Radio Frequency Sensor

The reflectance of the walnut leaves was collected with a radio frequency 3D sensor called Walabot Developer (Vayyar Imaging Ltd, Yehud-Monosson, Israel).[1] The frequency range was 3.3–10 GHz for the US/FCC model and 6.3–8 GHz for the EU/CE model [31]. The average transmitted energy of both models was below 41 dBm/MHz and did not constitute health concerns for the user, nor did it impact the tested tissues. Based on the technical specifications [31], the Walabot could sense the environment by transmitting, receiving, and recording signals from multiple antennas. Recordings of multiple transmit–receive antenna pairs were analyzed to build a 3D image of the examined environment. Changes in the environment were characterized by analyzing sequences of images. Then, researchers could detect changes in the environment by analyzing the sequences of images.

The Walabot used an antenna array to send radio frequency to the area in front of the walnut leaf and then captured the returning signals (reflectance or response). The exciting signals were produced, and the reflectance was then recorded by integrated circuits. The reflectance data were preprocessed by MATLAB 2020a. For example, records of the different reflectance strengths could be visualized in a graph (Fig. 7.2). The y-axis denoted the radio frequency reflectance from a walnut leaf. The x-axis meant the time. The research hypothesis was that the radio frequency reflectance of walnut leaves would be significantly different toward root-lesion nematode infection levels. Then, the spectral curve obtained from the leaves of walnut trees infected with nematodes could show the characteristic pattern of walnut under nematode stress.

7.2.3 Ground Truth Data Collection and Processing

In this chapter, the Walabot was used to measure the radio frequency reflectance of the walnut leaves in 2019 and 2020. For Project 45, there were 15 sampling trees from 6 different blocks in 2019 (Table 7.1) and 60 sampling trees from 4 different blocks in 2020 (Table 7.6). Each sampling tree was measured five times to reduce the likelihood of errors or anomalous results. Genotypes from one cross went into the analysis. "RLN per gram of root" meant the root-lesion nematode number per gram of root extracted in laboratory procedures. Based on the root-lesion nematode

[1] Mention of trade names or commercial products in this publication is solely to provide specific information and does not imply recommendation or endorsement by the University of California or the Kearney Agricultural Research and Extension Center. The University of California and Kearney Center are equal opportunity providers and employers.

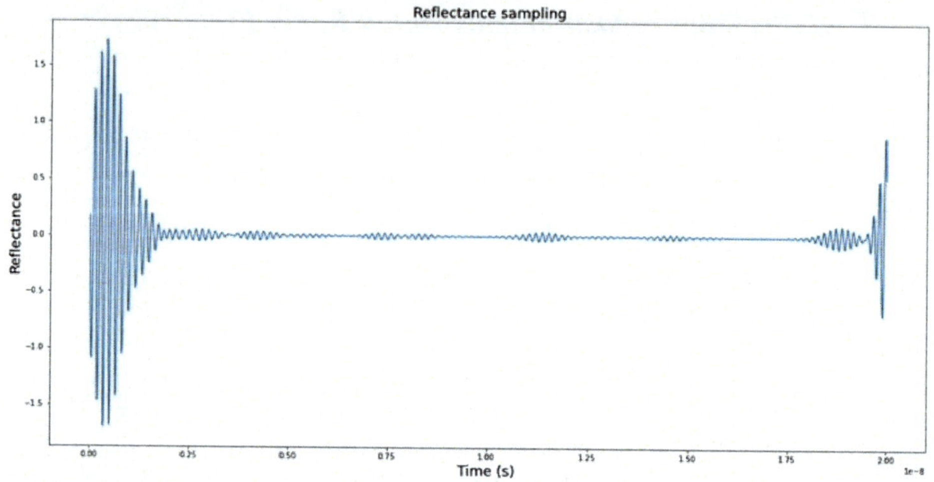

Fig. 7.2 The measurement signal of the radio frequencies reflectance of the walnut leaves using Walabot. The research hypothesis is that the reflectance of walnut leaves will be significantly different toward root-lesion nematode infection levels. The spectral curve obtained from the leaves of walnut trees infected with nematodes could show the characteristic pattern of walnut under nematode stress

Table 7.1 Nematode numbers per gram of root in rootstock genotypes examined by Walabot measurements (Project 45, 2019). "RLN per gram of root" meant the root-lesion nematode number per gram of root extracted in laboratory procedures

Block	Genotype	RLN per gram of root
A	62	57.2
A	95	49.8
A	127	70.6
A	117	99.8
B	72	147.2
B	89	8.0
B	127	47.7
C	95	50.2
C	127	40.8
D	48	72.6
D	127	7.1
D	128	117.6
E	89	76.1
E	127	1.8
F	127	0

Table 7.2 The classification of the nematode infestation levels (Project 45, 2019)

Nematode infestation levels	RLN per gram of root
Level 1	0–20
Level 2	20–60
Level 3	60–90
Level 4	>90

number, the walnut trees of 2019 were classified into four nematode infestation levels (Table 7.2).

7.2.4 Scikit-Learn Classification Algorithms

Several different methods in generating classifiers were adopted to evaluate the detection performance for nematode infection levels. “Neural Net,” “Support Vector Machine (SVM),” “Random Forest,” “AdaBoost,” “Nearest Neighbors,” and “Decision Tree” were chosen as the classification algorithms. In the “Neural Net” library, a multi-layer perceptron (MLP) classifier was used. This model optimized the log-loss function using stochastic gradient descent. MLP trained iteratively because the partial derivatives of the loss function with respect to the model parameters were computed to update the parameters at every step. The SVMs are a set of supervised learning methods used for classification, regression, and outlier detection. The SVMs are effective in high-dimensional spaces and effective in cases where the number of dimensions is greater than the number of samples.

For the “Random Forest” classifier, it is a meta-estimator that fits several decision tree classifiers on various sub-samples of the dataset and adopts averaging to improve the predictive accuracy and control overfitting. An “AdaBoost classifier” is also a meta-estimator that begins by fitting a classifier on the original dataset and then fits additional copies of the classifier on the same dataset but where the weights of incorrectly classified instances are adjusted such that subsequent classifiers focus more on complex cases.

The “Nearest Neighbors” method is to figure out a predefined number of training samples closest in the distance to the new point and predict the label from these. The number of samples can be a constant (k-nearest neighbor learning) or vary based on the local density of points (radius-based neighbor learning). Despite its simplicity, the nearest neighbors method has been successfully applied for many research problems, such as the handwritten digits classification. As a non-parametric method, it is often successful in classification situations where the decision boundary is very irregular.

“Decision Trees” are also non-parametric supervised learning methods commonly adopted for classification problems. The objective is to create a model that predicts the value of a target variable by learning simple decision rules inferred from the data features. A tree can be seen as a piecewise constant approximation.

"Decision Trees" usually uses a white box model, which means the explanation for the condition is easily explained by Boolean logic if a given situation is observable in a model. In contrast, results may be more challenging to interpret for a black box model, such as an artificial neural network.

7.2.5 *Deep Neural Networks (DNNs) and TensorFlow*

A deep neural network (DNN) usually has two or more hidden layers, commonly used in agriculture-related research. For example, in [1], Boniecki et al. adopted the DNN as one of the classification tools, with the process of identifying selected apple tree orchard pests as an example. In [36], You et al. proposed a segmentation method using DNN that could recognize the weed precisely with arbitrary shape in complex environmental conditions and enabled the autonomous robot to reduce the density of weed more successfully.

In this research, the authors trained a DNN with TensorFlow's high-level application programming interface (API). For each training instance of the nematode dataset, the algorithm fed the instance into the DNN and predicted the nematode infection level. The DNN classifier could be trained with any number of hidden layers. Then, a softmax output layer was used to generate the estimated nematode infection level probabilities. During the training process, the authors trained a DNN for classification with two hidden layers, one with 300 neurons and the other with 100 neurons.

7.3 Results and Discussion

7.3.1 *Data Visualization (Project 45, 2019)*

Figure 7.2 was a measurement example of the time-domain reflectance signal using the Walabot. Records of the different reflectance strengths could be visualized. Based on the root-lesion nematode number, the walnut trees were classified into four levels of nematode infestation (Table 7.2). Each sampling was 2×2048 dimension by default. Then, the collected nematode infestation data were converted into 2048-dimensional vectors for data processing. For data visualization, the authors applied Linear Discriminant Analysis (LDA) for dimension reduction, which learned the most critical axes between the classes [11]. The axes were then used to define hyperplane to project the high-dimensional training data into two dimensions, which gained important insight by visually detecting patterns (Fig. 7.3). The x-axis and y-axis had no scale because of hyperplane projection [7]. The nematode infection levels were successfully clustered into four levels. Although the two clusters of

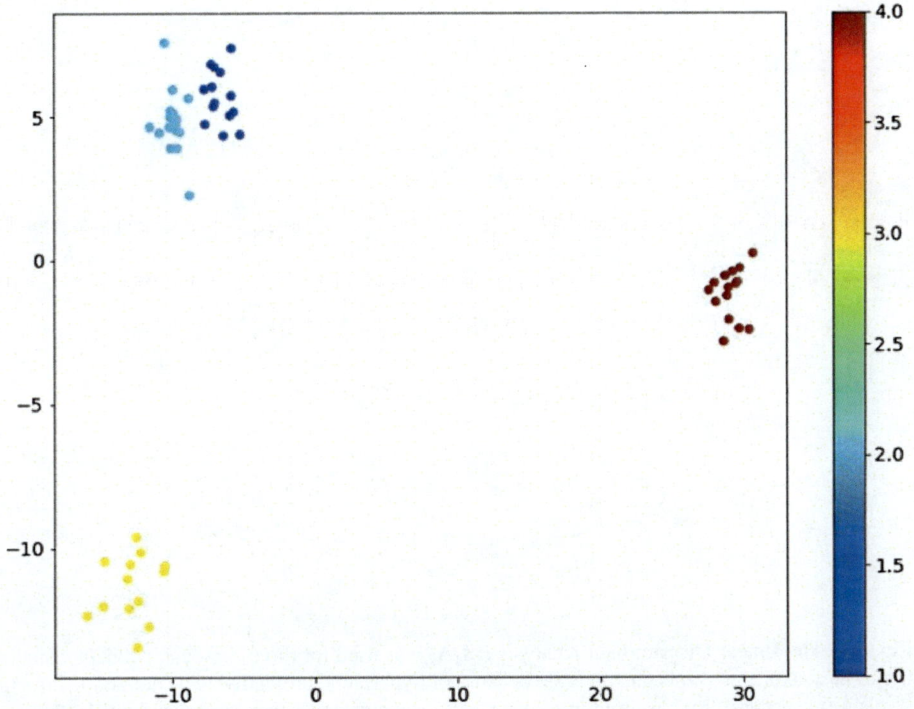

Fig. 7.3 The Linear Discriminant Analysis (LDA) was used for data visualization, which learned the most critical axes between the classes. The axes were then used to define hyperplane to project the high-dimensional training data into two dimensions, which gained important insight by visually detecting patterns. The x-axis and y-axis had no scale because of hyperplane projection. The nematode infection levels were successfully clustered into four levels. The color bar values 1, 2, 3, and 4 represented the four nematode infection levels from 1 to 4 (low to high) (Table 7.2)

lower nematode infection levels (Level 1 and Level 2) were close to each other in Fig. 7.3, they were clustered very well in three-dimensional space (Fig. 7.4).

7.3.2 Performance of Classifiers (Project 45, 2019)

The collected data of nematode infestation levels were converted into 2048-dimensional vectors for scikit-learn algorithms data processing. The data were distributed as 75% for training and 25% for testing using the *train_test_split* method. For evaluating the trained models, a confusion matrix was used to compare the performances of different classifiers. A confusion matrix was a summary of prediction results on a classification problem. The number of correct and incorrect predictions was tallied with count values and divided into classes. The confusion matrix provided insight into not only the errors being made by a classifier but also, more importantly, the types of errors that were being made. "True label" meant

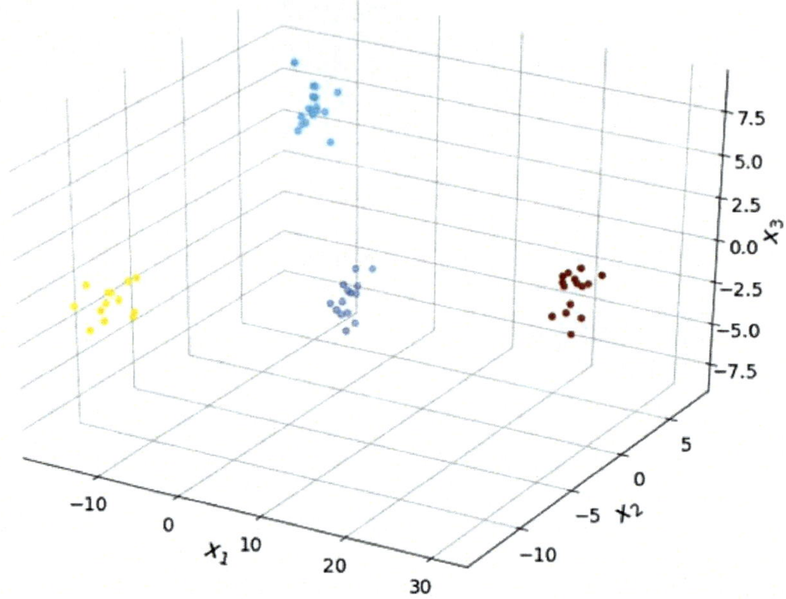

Fig. 7.4 The Linear Discriminant Analysis (LDA) was used for three-dimensional data visualization. The *x*-axis and *y*-axis had no scale because of hyperplane projection. The nematode infection levels were successfully clustered into four levels. The clusters' color was related to Fig. 7.3

the ground truth of nematode infestation levels. "Predicted label" identified the nematode infestation levels predicted by the trained model.

The trained scikit-learn classifiers had distinct test performance for early detection of nematode infection levels (Fig. 7.5 and Table 7.3). The *k*-nearest neighbors vote had the highest accuracy of 0.95. Table 7.4 shows the details of the test process of the *k*-nearest neighbors method, which was a type of instance-based learning that did not attempt to construct a general internal model, but simply stored instances of the nematode infection levels training data. The classification was computed from a simple majority vote of the nearest neighbors of each point. Then, a query point was assigned the data class that had the most representatives within the nearest neighbors of the point. For the other classifiers' test performance, the accuracy of the "Support Vector Machine" was 0.57. The "Random Forest" had a test accuracy of 0.76. The "AdaBoost," "Neural Network," and "Decision Tree" had an accuracy of 0.57, 0.62, and 0.81, respectively.

During the training process of DNN with TensorFlow, the number of epochs was set as 200. The batch size was 10 since there was a small training dataset. As shown in Fig. 7.6, there were 100 models. The highest accuracy could be as high as 0.9. The lowest accuracy was around 0.58 when the training instances could not represent the nematode infection levels very well. To evaluate the trained model, the authors run the training process 100 times to reduce the randomness of the training instances selection. The mean accuracy was 0.82 for the DNN classier model.

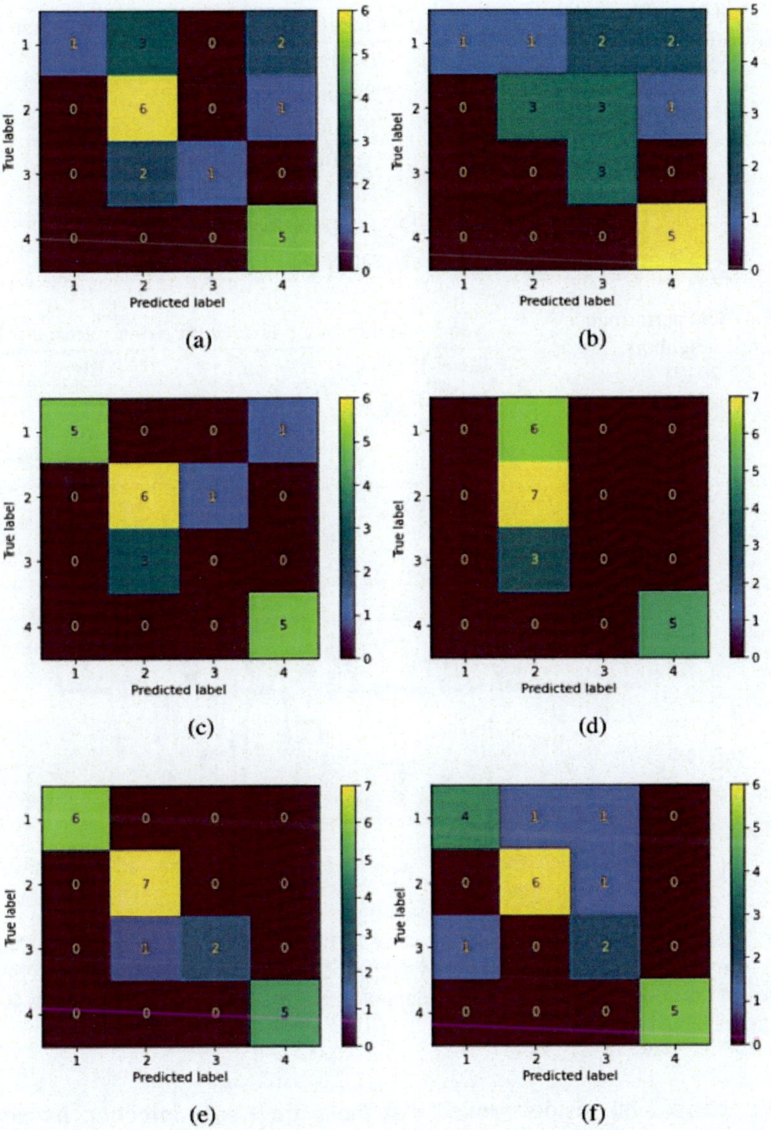

Fig. 7.5 A comparison of six classifiers in scikit-learn on nematode testing dataset (Project 45, 2019). A confusion matrix was a summary of prediction results on a classification problem. The number of correct and incorrect predictions was tallied with count values and divided into classes. The confusion matrix provided an insight into not only the errors being made by a classifier but more importantly the types of errors that were being made as well. "True label" meant the ground truth of nematode infestation levels. "Predicted label" identified the nematode infestation levels predicted by the trained model. (**a**) "Neural Net." (**b**) "Support Vector Machine." (**c**) "Random Forest." (**d**) "AdaBoost." (**e**) "Nearest Neighbors." (**f**) "Decision Trees"

Table 7.3 The performance of classification methods (Project 45, 2019)

Classification methods	Prediction accuracy
"Neural Network"	0.62
"Support Vector Machine"	0.57
"Random Forest"	0.76
"AdaBoost"	0.57
"Nearest Neighbors"	**0.95**
"Decision Tree"	0.81
"DNN with TensorFlow"	0.82

Table 7.4 The performance of *k*-nearest neighbors (Project 45, 2019)

Nematode infestation level	Precision	Recall	F1-score
Level 1	1	1	1
Level 2	0.88	1	0.93
Level 3	1	0.67	0.80
Level 4	1	1	1
Accuracy	NA	NA	**0.95**
Macro avg	0.97	0.92	0.93
Weighted avg	0.96	0.95	0.95

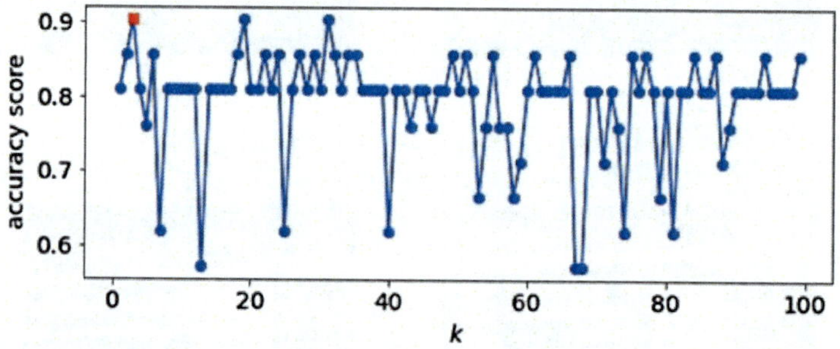

Fig. 7.6 There were k models (in this case, $k = 100$). The best accuracy could be as high as 0.9 (marked as a red dot). The lowest accuracy was around 0.58 when the training instances could not represent the nematode infection levels dataset very well. The mean accuracy was 0.82 for the DNN classier model

The experimental results proved that the walnut root infection by nematodes could induce spectral variation in walnut leaves. The proximate sensor could send the radio frequency to the area in front of the walnut leaf and then capture the returning signals (reflectance). Then, waveforms generated by different signals could estimate the damage caused by nematodes in the walnut root. With the machine learning algorithms, the nematode infection levels could be classified with a high accuracy of 95%. These findings emphasized the importance of applying phenotyping analysis for the early detection of nematode infection levels. Trees and crops severely infected by nematodes showed a significant reduction in leaf area similar to the plant response to other environmental stress, such as drought, nutrient deficiency, salinity, and other biotic stresses [13]. Several researchers reported the similar results [9, 18, 26, 27, 29]. For example, Matins et al. concluded that the multispectral classification with hyperspectral data could define the spatial distribution of healthy, moderately infected, and severely infected coffee plants, with an overall accuracy of 78% for detecting and mapping the root-knot nematode infection [18]. In [26], Palacharla et al. proposed an SVM regression method for the inversion and retrieval of key biophysical parameters that help understand and quantify the nature of the nematode infested vegetation. Therefore, to guarantee the efficiency of early detection of nematode infection levels, efficient machine learning algorithms and sensing technology should be adopted according to the results from this study.

7.3.3 Performance of Classifiers (Project 45, 2020)

In order to obtain a complete understanding of the facts and circumstances about the early detection of nematode infection levels, the authors also collected the data using the proximate sensor in 2020 (Table 7.5). For Project 45, there were 60 sampling trees from 4 different blocks in 2020 (Table 7.6). Each sampling tree was measured five times to reduce the likelihood of errors or anomalous results. Genotypes from one cross went into the analysis. Based on the root-lesion nematode number, the walnut trees of 2020 were classified into four levels of nematode infestation, from level 1 to level 4 (Table 7.5).

The machine learning algorithms were applied for nematode data processing and analysis. According to Fig. 7.7 and Table 7.7, the trained models had a different prediction performance for early detection of nematode infection levels. Although the exact prediction accuracy was distinct, the experimental results of 2020 were pretty consistent with the analysis results of 2019. The *k*-nearest neighbors vote still had the highest accuracy of 0.95. Table 7.8 shows the details of the test process of the *k*-nearest neighbors method. For the other classifiers' test performance, the accuracy of the "Support Vector Machine" was 0.58. The "Random Forest" had a test accuracy of 0.61. The "AdaBoost," "Neural Network," and "Decision Tree" had an accuracy of 0.63, 0.62, and 0.78, respectively.

For the training process of DNN with TensorFlow, the number of epochs was still set as 200 and the batch size was 10. As shown in Fig. 7.8, there were 100 models. The highest accuracy could be as high as 0.88. The lowest accuracy was around 0.4

Table 7.5 The classification of the nematode infestation levels (Project 45, 2020). "RLN per gram of root" meant the root-lesion nematode number per gram of root that was extracted in laboratory procedures

Nematode infestation levels	RLN per gram of root
Level 1	0–20
Level 2	20–60
Level 3	60–200
Level 4	>200

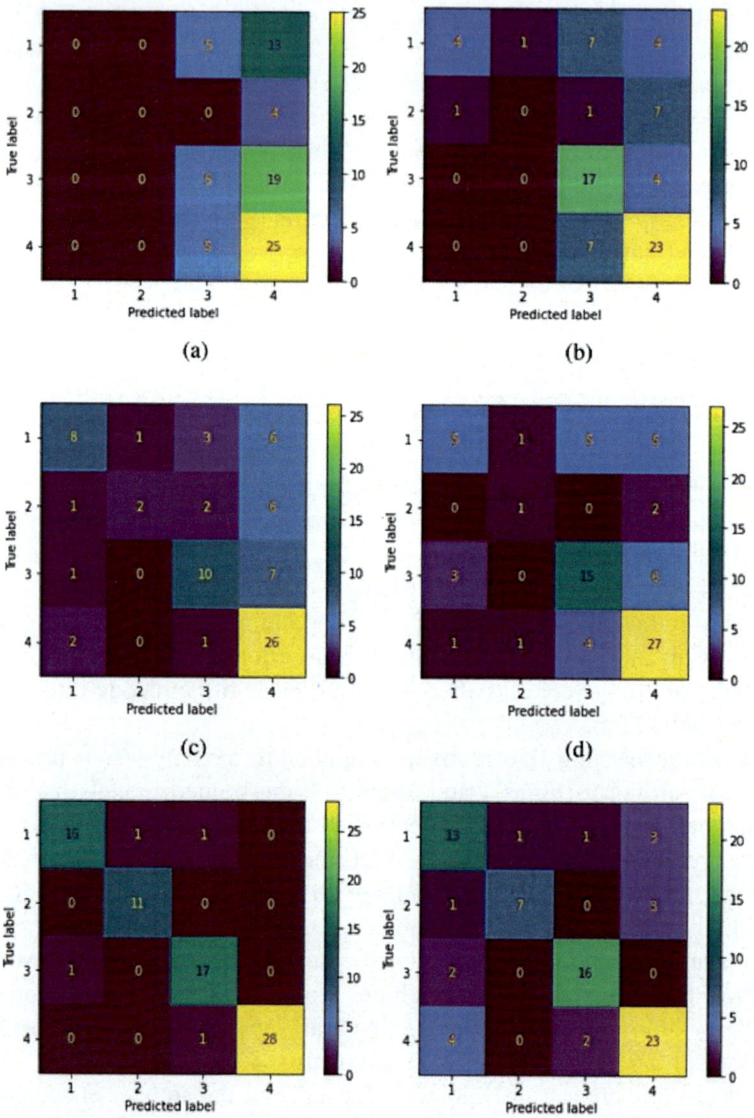

Fig. 7.7 A comparison of six classifiers in scikit-learn on nematode testing dataset (Project 45, 2020). Table 7.8 shows the details of the test process of the k-nearest neighbors method. (**a**) "Neural Net". (**b**) "Support Vector Machine". (**c**) "Random Forest". (**d**) "AdaBoost". (**e**) "Nearest Neighbors". (**f**) "Decision Trees"

Table 7.6 Nematode numbers per gram of root in rootstock genotypes examined by Walabot measurements (Project 45, 2020). "RLN per gram of root" meant the root-lesion nematode number per gram of root that was extracted in laboratory procedures. "ID" is a short form for "identity"

Tree ID	Genotype	RLN per gram of root	Tree ID	Genotype	RLN per gram of root
C1	MS1 48	243.8	A1	MS1 62	418.5
C2	I13-32	1005	A2	MS1 117	73.5
C3	Vlach	872.4	A3	MS1 117	43.5
C4	MS1 128	96	A4	MS1 95	205.5
C5	MS1 3	155.4	A5	MS1 48	82.5
C6	RX1	40.6	A6	MS1 128	355.5
C7	MS1 72	21	A7	Vlach	228
C8	MS1 117	4.5	A8	MS1 89	64.5
C9	MS1 31	207.6	A9	UZ229	172.5
C10	MS1 132	20	A10	MS1 127	37.5
C11	MS1 89	11.6	A11	MS1 3	1.4
C12	MS1 127	69	A12	RX1	1536
C13	MS1 95	201	A13	MS1 31	766.5
C14	VX211	19.5	A14	VX211	174
C15	MS1 62	462	A15	MS1 98	18
D1	Vlach	187.5	B1	MS1 31	67.5
D2	MS1 128	437.4	B2	MS1 48	111
D3	MS1 3	286.5	B3	MS1 95	57.4
D4	MS1 89	197.5	B4	MS1 127	15
D5	MS1 132	159	B5	MS1 132	23.2
D6	MS1 31	226.5	B6	VX211	36
D7	VX211	22.5	B7	MS1 62	141
D8	MS1 95	90	B8	Vlach	30
D9	MS1 127	114	B9	RX1	111
D10	MS1 72	931.5	B10	MS1 3	10.4
D11	MS1 122	201	B11	MS1 89	18
D12	I13-32	433.5	B12	MS1 72	27
D13	RX1	333	B13	MS1 117	12
D14	MS1 62	17.5	B14	UZ229	1090.5
D15	MS1 48	1599	B15	MS1 128	334.6

when the training instances could not represent the nematode infection levels very well. To evaluate the trained model, the authors run the training process 100 times to reduce the randomness of the training instances selection. The mean accuracy was around 0.7 for the DNN classier model.

As mentioned earlier in this section, the experimental results of 2020 were pretty consistent with the analysis results of 2019. Detecting the nematode disease in walnut roots with proximate sensing is usually based on changes in the spectral response of the walnut due to impairment of their leaf structure. The research

Table 7.7 The performance of classification methods (Project 45, 2020)

Classification methods	Prediction accuracy
"Neural Network"	0.63
"Support Vector Machine"	0.58
"Random Forest"	0.61
"AdaBoost"	0.63
"Nearest Neighbors"	**0.95**
"Decision Tree"	0.78
"DNN with TensorFlow"	0.7

Table 7.8 The performance of *k*-nearest neighbors (Project 45, 2020)

Nematode infestation level	Precision	Recall	F1-score
Level 1	0.94	0.89	0.91
Level 2	0.92	1	0.96
Level 3	0.89	0.94	0.92
Level 4	1	0.97	0.98
Accuracy	NA	NA	**0.95**
Macro avg	0.94	0.95	0.94
Weighted avg	0.95	0.95	0.95

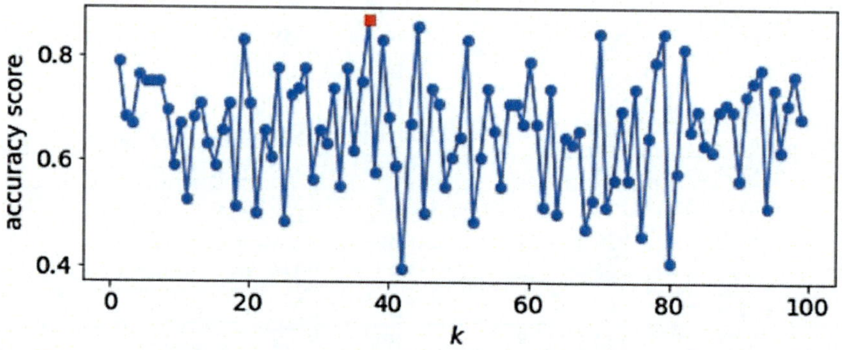

Fig. 7.8 The test performance of DNN for Project 45, 2020. There were 100 models and the best accuracy could be as high as 0.88 (marked as a red dot). The lowest accuracy was around 0.4 when the training instances could not represent the nematode infection levels dataset very well. The mean accuracy was 0.7 for the DNN classier model

consistency further enhanced the authors' discussion that the walnut root infection by nematodes could induce spectral variation in walnut leaves. The different signals generated by the radio frequency proximate sensor could estimate the walnut root damage caused by nematodes. With the *k*-nearest neighbors method, the nematode infection levels could be classified with a high accuracy of 95%. These findings further emphasized the importance of applying phenotyping analysis by machine learning algorithms for the early detection of nematode infection levels.

To evaluate the hypothesis that remote sensing is the best approach to monitoring root-knot nematodes in coffee crops, Matins et al. also collected agronomical parameters from plots of healthy and infected coffee plants, such as leaf area index (LAI), chlorophyll relative content obtained by soil and plant development analyzer (SPDA), biomass measurements, and radiometric data [18]. For the spectral characterization, radiometric data of leaves enabled the identification of the most sensitive spectral ranges for discrimination of the infected coffee plots. Inspired by the research work by Martins et al. [18], the authors will also collect agronomical parameters from the study site in the future and would further figure out the correlation between the radio frequency reflectance and the agronomical parameters, such as LAI and chlorophyll relative content of the walnut leaves.

7.4 Chapter Summary

The aim of the present chapter is the early detection of nematode infection levels in the walnut root by a non-invasive method. The authors collected the radio frequency reflectance data of walnut leaves by using a proximate sensor. Then, machine learning algorithms were adopted for the nematode infection levels classification problem. The research results showed that the best classification accuracy of nematode infection levels was 95% when the k-nearest neighbors method was being adopted. The walnut root infection by nematodes could induce spectral variation in walnut leaves. The results of this research supported the idea that trees and crops severely infected by nematodes showed a significant reduction in leaf area similar to the plant response to other environmental stress, such as drought, nutrient deficiency, salinity, and other biotic stresses. Furthermore, the findings of this research provided insights for the early detection of nematodes using phenotyping and machine learning technology.

The study was limited by the lack of information on the biophysical explanation of what variations of the walnut leaves were responsive or sensitive to the radio frequency signal under the pressure of nematodes. Despite its limitations, the study certainly added an understanding of the effect of nematode infection on walnut phenotyping analysis. In the future, the precise mechanism of the radio frequency sensor in the early detection of nematodes remains to be elucidated. The authors will also collect agronomical parameters from the study site and figure out the correlation between the radio frequency reflectance and the agronomical parameters, such as LAI and chlorophyll relative content of the walnut leaves. The authors will also perform similar data analysis for Project 30 (a, b) and Project Smith.

References

1. Boniecki, P., Zaborowicz, M., Pilarska, A., Piekarska-Boniecka, H.: Identification process of selected graphic features apple tree pests by neural models type MLP, RBF and DNN. Agriculture **10**(6), 218 (2020)
2. Breiman, L.: Random forests. Mach. Learn. **45**(1), 5–32 (2001)

3. Brito, J.A., Dickson, D.W., Kaur, R., Vau, S., Stanley, J.D.: The peach root-knot nematode: Meloidogyne floridensis, and its potential impact for the peach industry in Florida. Florida Department of Agriculture and Consumer Services (2015)
4. Chang, C.C., Lin, C.J.: LIBSVM: a library for support vector machines. ACM Trans. Intell. Syst. Technol. **2**(3), 1–27 (2011)
5. Eisenback, J.D., Triantaphyllou, H.H., et al.: Root-knot nematodes: Meloidogyne species and races. Manual Agric. Nematol. **1**, 191–274 (1991)
6. Friedman, J., Hastie, T., Tibshirani, R.: The Elements of Statistical Learning, vol. 1. Springer Series in Statistics, New York (2001)
7. Géron, A.: Hands-on Machine Learning with Scikit-Learn, Keras, and TensorFlow: Concepts, Tools, and Techniques to Build Intelligent Systems. O'Reilly Media (2019)
8. Goldberger, J., Hinton, G.E., Roweis, S., Salakhutdinov, R.R.: Neighbourhood components analysis. In: Advances in Neural Information Processing Systems, vol. 17, pp. 513–520 (2004)
9. Guerra, M., Paba, G., Cubillán, N., Portillo, E., Casassa-Padrón, A.M., Aballay, E., San-Blas, E.: Sequential spectral changes of Meloidogyne enterolobii-infected plants using two dimensional correlation IR spectroscopy. Nematology **1**, 1–15 (2020)
10. Hillnhütter, C., Mahlein, A.K., Sikora, R.A., Oerke, E.C.: Use of imaging spectroscopy to discriminate symptoms caused by Heterodera schachtii and Rhizoctonia solani on sugar beet. Precis. Agric. **13**(1), 17–32 (2012)
11. Izenman, A.J.: Linear discriminant analysis. In: Proceedings of the Modern Multivariate Statistical Techniques, pp. 237–280. Springer, Berlin (2013)
12. Jolliffe, I.: Principal Component Analysis. Springer, Berlin (2011)
13. Jones, H.G., Vaughan, R.A.: Remote Sensing of Vegetation: Principles, Techniques, and Applications. Oxford University Press, Oxford (2010)
14. Kingma, D.P., Ba, J.: Adam: a method for stochastic optimization (2014). arXiv preprint arXiv:1412.6980
15. Lawrence, G.W., King, R., Ellis, G., Doshi, R., Lawrence, K., Caceres, J., Samson, S.: Population estimation of the reniform nematode using hyperspectral reflectance data and applications to variable rate nematicide applications, vol. 1. In: Proceedings of the National Beltwide Cotton Conference, National Cotton Council, Memphis (2007)
16. Li, R., Lu, Y., Peters, J.M., Choat, B., Lee, A.J.: Non-invasive measurement of leaf water content and pressure–volume curves using terahertz radiation. Sci. Rep. **10**(1), 1–14 (2020)
17. Loh, W.Y.: Classification and regression trees. Wiley Interdiscip. Rev. Data Mining Knowl. Discov. **1**(1), 14–23 (2011)
18. Martins, G.D., Galo, M.d.L.B.T., Vieira, B.S.: Detecting and mapping root-knot nematode infection in coffee crop using remote sensing measurements. IEEE J. Sel. Topics Appl. Earth Obs. Remote Sensing **10**(12), 5395–5403 (2017)
19. McKenry, M., Buzo, T., Kaku, S., McKenna, J.: A three-year search for nematode resistance in walnut-final report. Walnut Res. Rep. **2000**, 509–512 (2001)
20. McKenry, M.V.: The replant problem and its management (1999). http://kare.ucanr.edu/programs/Nematodes/The_Replant_Problem/index.cfm?editon=0. Accessed 19 Mar 2020
21. Niu, H., Hollenbeck, D., Zhao, T., Wang, D., Chen, Y.: Evapotranspiration estimation with small UAVs in precision agriculture. Sensors **20**(22), 6427 (2020)
22. Niu, H., Wang, D., Chen, Y.: Estimating actual crop evapotranspiration using deep stochastic configuration networks model and UAV-based crop coefficients in a pomegranate orchard. In: Autonomous Air and Ground Sensing Systems for Agricultural Optimization and Phenotyping V. International Society for Optics and Photonics (2020)
23. Niu, H., Zhao, T., Wang, D., Chen, Y.: Estimating evapotranspiration with UAVs in agriculture: a review. In: Proceedings of the ASABE Annual International Meeting. American Society of Agricultural and Biological Engineers (2019)
24. Niu, H., Zhao, T., Wang, D., Chen, Y.: A UAV resolution and waveband aware path planning for onion irrigation treatments inference. In: 2019 International Conference on Unmanned Aircraft Systems (ICUAS), pp. 808–812. IEEE, Piscataway (2019)

25. Niu, H., Zhao, T., Westphal, A., Chen, Y.: A low-cost proximate sensing method for early detection of nematodes in walnut using Walabot and scikit-learn classification algorithms. In: Autonomous Air and Ground Sensing Systems for Agricultural Optimization and Phenotyping V. International Society for Optics and Photonics (2020)
26. Palacharla, P.K., Durbha, S.S., King, R.L., Gokaraju, B., Lawrence, G.W.: A hyperspectral reflectance data based model inversion methodology to detect reniform nematodes in cotton. In: Proceedings of the 6th International Workshop on the Analysis of Multi-temporal Remote Sensing Images (Multi-Temp), pp. 249–252. IEEE, Piscataway (2011)
27. Riefolo, C., Antelmi, I., Castrignanò, A., Ruggieri, S., Galeone, C., Belmonte, A., Muolo, M.R., Ranieri, N.A., Labarile, R., Gadaleta, G., et al.: Assessment of the hyperspectral data analysis as a tool to diagnose Xylella fastidiosa in the asymptomatic leaves of olive plants. Plants **10**(4), 683 (2021)
28. Rosenblatt, F.: The perceptron: a probabilistic model for information storage and organization in the brain. Psychol. Rev. **65**(6), 386 (1958)
29. San-Blas, E., Paba, G., Cubillán, N., Portillo, E., Casassa-Padrón, A.M., González-González, C., Guerra, M.: The use of infrared spectroscopy and machine learning tools for detection of Meloidogyne infestations. Plant Pathol. **69**(8), 1589–1600 (2020)
30. Singh, S., Singh, B., Singh, A.: Nematodes: a threat to sustainability of agriculture. Proc. Environ. Sci. **29**, 215–216 (2015)
31. Walabot: Walabot Technical Brief (2018). https://walabot.com/docs/walabot-tech-brief-416?type=pdf
32. Wang, Y., Chen, Y., Liao, X., Dong, L.: Lithium-ion battery face imaging with contactless Walabot and machine learning. In: Proceedings of the 2019 IEEE International Conference on Mechatronics and Automation (ICMA), pp. 1067–1072. IEEE, Piscataway (2019)
33. Wang, Y., Niu, H., Zhao, T., Liao, X., Dong, L., Chen, Y.: Contactless Li-Ion battery voltage detection by using Walabot and machine learning. In: Proceedings of the ASME 2019 International Design Engineering Technical Conferences and Computers and Information in Engineering Conference. American Society of Mechanical Engineers Digital Collection (2019)
34. Westerdahl, B., Westphal, A., Hasey, J.K., McKenry, M.: UC IPM pest management guidelines: Walnut nematodes (2017). https://www2.ipm.ucanr.edu/agriculture/walnut/Nematodes/. Accessed 19 Mar 2020
35. Westphal, A., Maung, Z.T., Doll, D.A., Yaghmour, M.A., Chitambar, J.J., Subbotin, S.A.: First report of the peach root-knot nematode, Meloidogyne floridensis infecting almond on root-knot nematode resistant 'Hansen 536' and 'Bright's hybrid 5'rootstocks in California, USA. Society of Nematologists (2019)
36. You, J., Liu, W., Lee, J.: A DNN-based semantic segmentation for detecting weed and crop. Comput. Electron. Agric. **178**, 105,750 (2020)
37. Zhao, T., Koumis, A., Niu, H., Wang, D., Chen, Y.: Onion irrigation treatment inference using a low-cost hyperspectral scanner. In: Multispectral, Hyperspectral, and Ultraspectral Remote Sensing Technology, Techniques and Applications VII. International Society for Optics and Photonics (2018)

Chapter 8
Tree-Level Evapotranspiration Estimation of Pomegranate Trees Using Lysimeter and UAV Multispectral Imagery

Abstract This chapter presents an in-depth investigation into the tree-level evapotranspiration estimation of pomegranate trees, employing a novel approach that integrates lysimeter measurements with Unmanned Aerial Vehicle (UAV) multispectral imagery. The chapter begins with an introduction, underscoring the importance of accurate evapotranspiration estimation for effective water resource management in pomegranate orchards. The materials and methods are meticulously detailed, including the study site description and the process of UAV image collection and processing. The methodology includes the determination of individual tree crop coefficients (K_c) from Normalized Difference Vegetation Index (NDVI), spatial variability mapping of K_c and crop evapotranspiration (ET_c), and the assessment of performance in individual tree-level evapotranspiration estimation. Results and discussions unfold, revealing insights into the determination of individual tree K_c, the spatial variability mapping of K_c and ET_c, and the overall performance of the individual tree-level evapotranspiration estimation method. The chapter concludes with a set of comprehensive conclusions, summarizing key findings and their implications for advancing digital agriculture practices in pomegranate cultivation.

8.1 Introduction

Evapotranspiration (ET) estimation is important for digital agriculture, especially precision water management. The temporal and spatial mapping of ET can identify variations in the field, which is useful for evaluating soil moisture [45, 48] and assessing crop water status [18]. ET estimation can also benefit water resource management and weather forecast [47]. ET is a combination of two separate processes, evaporation (E) and transpiration (T). Evaporation is the process whereby liquid water is converted to water vapor through latent heat exchange [2]. Transpiration is the process of the vaporization of liquid water contained in plant tissues and the removal of vapor from the atmosphere [2]. The current theory of transpiration is composed of the following three steps. First, the conversion of liquid-phase water to vapor water causes canopy cooling from latent heat exchange. Therefore, canopy temperature can be used as an indicator of ET. Second, diffusion of water vapor

H. Niu, Y. Chen, *Smart Big Data in Digital Agriculture Applications*, Agriculture Automation and Control, https://doi.org/10.1007/978-3-031-52645-9_8

from inside the plant stomata to the surrounding atmosphere. Third, atmospheric air mixing by convection or diffusion transports vapor near the plant surfaces to the upper atmosphere or off-site away from the plant canopy. Usually, evaporation and transpiration occur simultaneously.

Many approaches have been developed to estimate ET. Typically, there are direct and indirect methods. For direct methods, ET can be determined by water balance [50]:

$$ET = P + I - D - R - S, \tag{8.1}$$

where P (mm day^{-1}) is precipitation, I (mm day^{-1}) is irrigation, D (mm day^{-1}) is drainage, R (mm day^{-1}) is runoff, and S (mm day^{-1}) is the soil moisture storage. These direct ET methods, however, are usually point-specific or area-weighted measurements and cannot be extended to a large scale because of the heterogeneity of the land surface. The experimental equipment is also costly and requires substantial expense and effort, such as lysimeters, which are only available for a small group of researchers. For indirect methods, there are energy balance methods [22] and remote sensing methods [1]. For energy balance methods, Bowen ratio [5, 13] and eddy covariance [27] have been widely used in ET estimation. However, they are also area-weighted measurements. Remote sensing techniques can detect variations in vegetation and soil conditions over space and time. Thus, they have been considered as one of the most powerful methods for mapping and estimating spatial ET over the past decades [19, 42]. Remote sensing models have been useful in accounting for the spatial variability of ET at regional scales when using satellite platforms such as Landsat and ASTER [3, 6, 21, 38]. Since the satellite was being applied [23], several remote sensing models have been developed to estimate ET, such as the Surface Energy Balance Algorithm for Land (SEBAL) [1, 6], Mapping Evapotranspiration with Internalized Calibration (METRIC) [4], the Dual Temperature Difference (DTD) [36], and the Priestley–Taylor TSEB (TSEB-PT) [15]. Remote sensing techniques can provide information such as Normalized Difference Vegetation Index (NDVI), Leaf Area Index (LAI), surface temperature, and surface albedo. Related research on these parameters has been discussed by different researchers [20, 26, 40].

As a new remote sensing platform, researchers are increasingly interested in the potential of small UAVs in digital agriculture [9, 14, 43, 51], especially in heterogeneous crops, such as vineyards and orchards [54, 56]. UAVs overcome some of the remote sensing limitations faced by satellites. For example, satellite remote sensing is prone to cloud cover; UAVs are below the clouds. Compared to satellites, UAVs can be operated at any time if the weather is within operating limits. The satellite has a fixed flight path, and UAVs are more mobile and adaptive for site selection. Mounted on the UAVs, lightweight sensors, such as RGB cameras, multispectral cameras, and thermal infrared cameras, can be used to collect high-resolution images. The higher temporal and spatial resolution images, relatively low operational costs, and the nearly real-time image acquisition make UAVs an ideal platform for mapping and monitoring ET. Many researchers have already used UAVs and lightweight sensors for ET estimation, as shown in Tables 8.1 and 8.2. For example, in [37], Ortega-Farías *et al.* implemented a remote sensing energy balance (RSEB) algorithm for estimating energy components in an olive orchard, such as incoming solar radiation, sensible heat flux, soil heat flux, and latent heat flux. Optical sensors were mounted on a UAV to provide high spatial resolution images. By using the UAV platform, experiment results show that the RSEB algorithm can estimate latent heat flux and sensible heat flux with errors of 7% and 5%, respectively. It demonstrated that UAV could be used as an excellent platform to evaluate the spatial variability of ET in the olive orchard.

Because of the recurring water shortages in California, many growers started growing crops that have drought resistance and high economic value to a certain degree [53], such as pomegranate. There is around 11,000 ha of pomegranate in California [52]. Evidence suggests that evapotranspiration (ET) estimation is among the most important factors to manage limited water effectively in agriculture [2]. Mapping the ET temporally and spatially can identify variations in the field, which is useful for evaluating soil moisture [45, 48] and assessing crop water status [18].

Using crop coefficient (K_c) to estimate ET is a common method for water irrigation management. The actual crop evapotranspiration (ET_c) is determined by the K_c approach, whereby the effect of the various weather conditions is incorporated into reference ET (ET_o) and the crop characteristics into the K_c [2]. At the beginning of the growing season, K_c increases from a minimum value. When the canopy cover is full, the K_c reaches a maximum around the mid-season. Then, the K_c starts decreasing before the end of the growing season. The normalized difference vegetation index (NDVI) has been commonly used for vegetation monitoring, such as water stress detection [55, 57], crop yield assessment [58], and ET estimation [31–35]. The value of NDVI is a standardized method to measure healthy vegetation. When the NDVI is high, it indicates the vegetation has a higher level of photosynthesis. Many studies have used satellite-derived NDVI to estimate crop coefficient values [16, 17, 41]. For example, Trout *et al.* [44] and Zhang *et al.* [52] applied a remote sensing method using NDVI to estimate canopy ground cover as a basis for generating K_c. Kamble *et al.* [17] used a simple linear regression model to establish a relationship between NDVI and K_c. Although satellite imagery can provide spatially distributed measurements, it cannot acquire useful spatio-temporal resolution images for digital agriculture applications [12]. The timing of satellite overpass is not always synchronous with research requirements, either. To date, few studies have investigated the association between the K_c and NDVI at the individual tree level.

The **objectives of this chapter** were (1) to investigate and validate the approaches of estimating K_c using UAV-based NDVI for an experimental pomegranate orchard, (2) to establish a linear regression model between the NDVI and K_c in the individual tree level, and (3) to evaluate the performance of the new regression model on estimating 100% ET irrigation sampling trees. The **major contributions of this chapter** were (1) developed a reliable tree-level ET estimation method using UAV high-resolution multispectral images and (2) provided a framework to establish a linear regression model between the NDVI and

Table 8.1 ET estimation using UAV platforms

Study sites	UAV platforms	Sensors	Method	Crops	References
Ames, IA, USA	eBee Ag	Sequoia, Cannon S110 ThermoMAP camera	SEBAL	Corn and soybean	[25]
Scipio, UT Lodi, CA, USA	AggieAir	Canno S-95 ICI thermal camera	METRIC	Vineyard	[11]
Pinto Bandeira city Rio Grande do Sul State, Brazil	AIBOTIX	Nikon CoolpixA Camera	METRIC	Vineyard	[10]
HOBE agricultural site, Denmark	Q300, QuestUAV	Optris PI 450	TSEB	Barley	[15]
Lodi, CA, USA	Cessna TU206	ImperX Bobcat B8430 ThermaCAM SC640	TSEB	Vineyard	[49]
Lodi, CA, USA	AggieAir	NA	TSEB	Vineyard	[28]
Pinto Bandeira Serra Gaucha, Brazil	AIBOTIX Hexacoptero	Nikon CoolpixA camera	TSEB	Vineyard	[24]
Lodi, CA, USA	NA	NA	TSEB	Vineyard	[29]
Pencahue Valley Región del Maule, Chile	NA	Mini MCA-6 EasIR-9	TSEB	Olive	[37]
Bushland, TX, USA	AggieAir	Kodak1 thermal infrared model 760	TSEB	Sorghum and corn	[8]
Petit-Nobressart, Luxembourg	MikroKopter OktoXL	Samsung ES80 Optris Pi 400	TSEB	Grassland	[7]
Lodi, CA, USA	Cessna TU206	ImperX Bobcat B8430 ThermaCAM SC640	OSEB	Vineyard	[49]
Petit-Nobressart, Luxembourg	MikroKopter OktoXL	Samsung ES80 Optris Pi 400	OSEB	Grassland	[7]
Tatura, Victoria, Australia	DJI S1000	A65 and RedEdge M	HRMET	Peach, nectarine, and corn	[38]

Table 8.2 Multispectral and thermal infrared sensors on UAV platforms

Sensor	Function	Resolution	Weights	Dimensions	Spectral bands	Accuracy
RedEdge M	Multispectral	1280×960 pixels	231.9 g	8.7×5.9×4.54 cm	Blue, green, red, red edge, and near-infrared (NIR)	8.2 cm/pixel, per band at 120 m
MAPIR Survey 3	Multispectral	4608×3456 pixels	76 g	5.9×4.15×3.6 cm	375–650 nm	4.05 cm/pixel at 120 m
Mini MCA-6	Multispectral	1280×1024 pixels	700 g	13.14×7.83×8.76 cm	450–1000 nm	3.3 cm/pixel at 60 m
Tetracam ADC Lite	Multispectral	2048×1536 pixels	200 g	11.4×7.7×6.05 cm	Red, green, NIR	5 cm/pixel at 150 m
Sequoia	Multispectral	4608×3456 pixels	72 g	5.9×4.1×2.8 cm	Green, red, red edge, and near-infrared	17 cm/pixel at 100 m
Cannon S 110	Near-infrared	4000×3000 pixels	198 g	9.9×5.9×2.7 cm	NIR	3.5 cm/pixel at 100 m
ICI SWIR 640 P	Short-wave infrared	640×512 pixel	130 g	4.6×4.6×2.95 cm	0.9–1.7 μm	±1 °C
ICI 9640 P	Thermal infrared	640×480 pixel	37 g	3.4×3.0×3.4 cm	7–14 μm	±1 °C
ICI 8640 P	Thermal infrared	640×480 pixel	74.5 g	4.5×4.5×3.9 cm	7–14 μm	±1 °C
FLIR Vue Pro R 640	Thermal infrared	640×512 pixel	72 g	5.74×4.44 cm	7.5–13.5 μm	±5 °C
Optris PI 450	Thermal infrared	382×288 pixels	240 g	4.6×5.6×6.8 cm	8–14 μm	±2 °C or ±2%
ThermalCAM SC640	Thermal infrared	640×480 pixel	1.7 kg	28.2×14.4×14.7 cm	7.5–13 μm	±2 °C or ±2%
EasIR-9	Thermal infrared	288×384 pixel	1 kg	11.2×18.2×25.2 cm	8–14 μm	±2 °C or ±2%
ThermoMAP	Thermal infrared	640×512 pixel	134 g	56×46×26 cm	7–15 μm	±5 °C
A65	Thermal infrared	640×512 pixel	200 g	29.5×20.0×10.5 cm	7.5–13 μm	±5 °C
Optris PI 400	Thermal infrared	382×288 pixel	240 g	4.6×5.6×6.8 cm	8–14 μm	±2 °C

K_c to estimate the actual daily ET. Results showed that the linear regression model could estimate tree-level ET with an R^2 and mean absolute error (MAE) of 0.9143 and 0.39 mm/day, respectively, which showed a state-of-the-art performance.

8.2 Materials and Methods

8.2.1 Study Site Description

Field studies were conducted in an experimental pomegranate (*Punica granatum L.*, cv "Wonderful") field (Fig. 8.1) at the USDA-ARS, San Joaquin Valley Agricultural Sciences Center (36.594°N, 119.512°W), Parlier, CA, USA. The San Joaquin Valley has a Mediterranean climate with hot and dry summers. Rainfall is insignificant, and irrigation is the only source of water for pomegranate growth [46]. There are two large weighing lysimeters installed at the center of the experimental field [39]. According to [53], the lysimeters have a resolution of approximately 0.1 mm of water loss. The soil types are a Hanford fine sandy loam (coarse-loamy, mixed, and thermic Typic Xerorthents). The meteorological data were generated by CIMIS (California Irrigation Management Information System) weather station 39, which is about 700 m far from the experimental field.

The pomegranate field was randomly designed into 16 equal blocks, with four replications, to test four irrigation levels. The irrigation volumes are 35%, 50%, 75%, and 100% of ET_c, which was measured by the weighing lysimeter in the field.

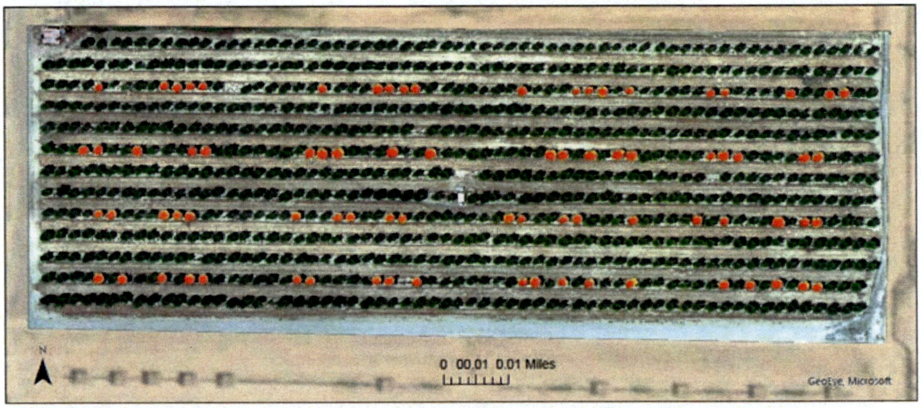

Fig. 8.1 The pomegranate study site. Field studies were conducted in an experimental pomegranate (*Punica granatum L.*, cv "Wonderful") field at the USDA-ARS, San Joaquin Valley Agricultural Sciences Center (36.594 °N, 119.512 °W), Parlier, CA, USA. The pomegranate field was randomly designed into 16 equal blocks, with four replications, to test four irrigation levels. The irrigation volumes are 35%, 50%, 75%, and 100% of ET_c, which was measured by the weighing lysimeter in the field. There were five sampling trees in each block, 80 sampling trees in total, marked with red labels

There were five sampling trees in each block, 80 sampling trees in total, marked with red labels in Fig. 8.1.

8.2.2 UAV Image Collection and Processing

A UAV platform, called the Hover, was deployed for imagery data acquisition. The RedEdge M (MicaSense, Seattle, WA, USA) was used for collecting multispectral images. The RedEdge M has five different bands, which are Blue (B, 475 nm), Green (G, 560 nm), Red (R, 668 nm), Red Edge (RE, 717 nm), and Near-Infrared (NIR, 840 nm). With a Downwelling Light Sensor (DLS), a five-band light sensor that connects to the camera, the RedEdge M can measure the ambient light during a flight mission for each of the five bands. Then, it can record the light information in the metadata of the images captured by the camera. After the camera calibration, the information detected by the DLS can be used to correct lighting changes during a flight, such as changes in cloud cover during a UAV flight.

The UAV flight missions were configured by using the MissionPlanner (ArduPilot, USA). The flight altitude was set up as 60 m. The overlapping of UAV images was set up as 80% forward and 70% by the side. Then, the UAV imagery of the pomegranate can be stitched together during image processing with high confidence. The image of a calibrated reflectance panel (CRP) was taken before and after the flight missions, servicing as the reflectance reference. The authors flew the UAV bi-weekly over the pomegranate field at noon during the growing season in 2019. The successful data collections were shown in Table 8.3. After the flight missions, all aerial images were stitched together to generate the orthomosaic images in PhotoScan (Agisoft LLC, Russia).

Table 8.3 Flight missions at the USDA in 2019. The UAV flight missions were configured by using the MissionPlanner (ArduPilot, USA). The flight altitude was set up as 60 m. The overlapping of UAV images was set up as 80% forward and 70% by the side

Dates	Flight time	Flight altitude
May 8, 2019	12–1 pm	60 m
July 25, 2019	12–1 pm	60 m
August 7, 2019	12–1 pm	60 m
August 29, 2019	12–1 pm	60 m
September 19, 2019	12–1 pm	60 m
October 3, 2019	12–1 pm	60 m
October 29, 2019	12–1 pm	60 m

8.3 Results and Discussion

8.3.1 Determination of Individual Tree K_c from NDVI

The correlation between the K_c and NDVI of an individual tree was analyzed. Daily K_c for the individual tree was calculated as [2]

$$K_c = ET_c/ET_o, \tag{8.2}$$

where the actual ET (ET_c) was measured by the weighing lysimeter and the reference ET (ET_o) was obtained from the CIMIS weather station near the field. The mean NDVI values for the lysimeter tree were calculated by ρ_{NIR} and ρ_R, where ρ_{NIR} and ρ_R were the reflectance of near-infrared and red wavebands, respectively. NDVI is a standardized method to measure healthy vegetation. When the NDVI is high, it indicates the vegetation has a higher level of photosynthesis (one of our demos is an NDVI mapping for May 8, 2019). According to the demo, most NDVI values of the tree canopies range from 0.468 to 1. It is interesting to point out that the shade of the trees had a mean NDVI value around 0.5.

$$NDVI = \frac{\rho_{NIR} - \rho_R}{\rho_{NIR} + \rho_R}. \tag{8.3}$$

Key Observation Figure 8.2 showed the relationship between the daily K_c and the derived mean NDVI for the tree in the lysimeter with 100% irrigation treatment. The sampling data started on May 8 and ended on October 29. The linear relationship had an intercept of 0.6114 and a slope of 1.6493. It also had a high correlation coefficient of 0.8865, indicating a significant correlation between the K_c and NDVI at the individual tree level during the growing season in 2019.

8.3.2 The Spatial Variability Mapping of K_c and ET_c

Key Observation Inspired by the strong linear correlation between the K_c and NDVI for the individual lysimeter tree, the authors calculated the K_c for all the trees in the experimental field using the linear regression model. The mapping of K_c and ET_c was generated, respectively (Fig. 8.3 is one example for K_c and ET_c spatial variability mapping on May 8). The values of K_c were mostly between 0.578 and 1.039 on May 8. For the ET_c, it ranged from 3.2 to 6.0 mm/day.

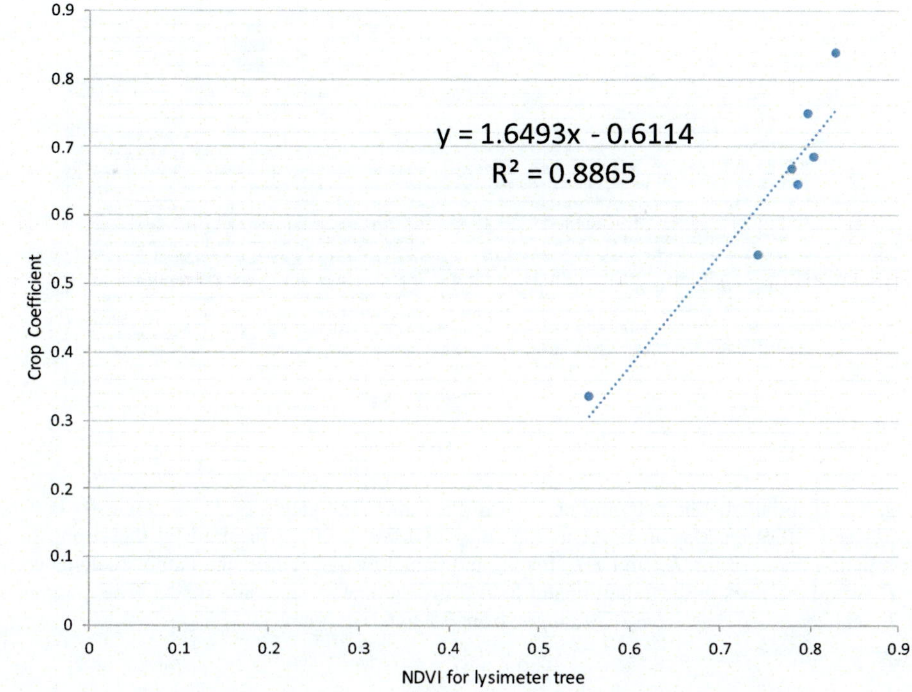

Fig. 8.2 Seasonal K_c and NDVI for the tree in lysimeter. The sampling data started on May 8 and ended on October 29. The linear relationship had an intercept of 0.6114 and a slope of 1.6493. It also had a high correlation coefficient of 0.8865, indicating a significant correlation between the K_c and NDVI at the individual tree level during the growing season in 2019

(a)

Fig. 8.3 The K_c and ET_c maps of May 8, 2019. (**a**) The K_c map of May 8, 2019. Inspired by the strong linear correlation between the K_c and NDVI for the individual lysimeter tree, the authors calculated the K_c for all the trees in the experimental field using the linear regression model.

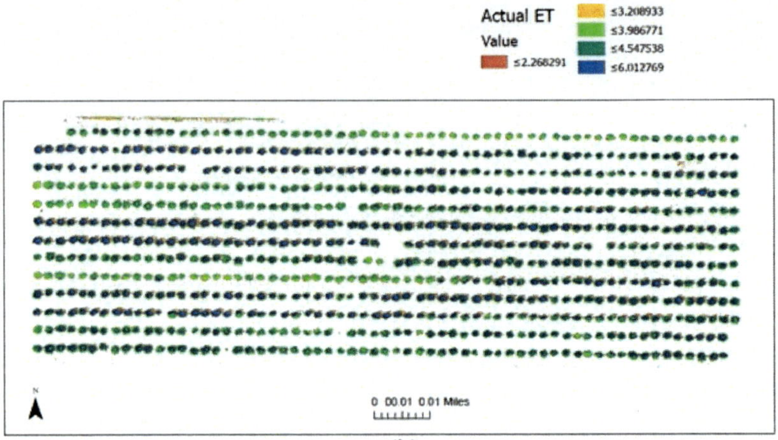

(b)

Fig. 8.3 (continued) The mapping of K_c was generated. The values of K_c were mostly between 0.578 and 1.039 on May 8. (**b**) The ET_c map of May 8, 2019. Inspired by the strong linear correlation between the K_c and ET_c for the individual lysimeter tree, the authors calculated the ET_c for all the trees in the experimental field using the linear regression model. The mapping of ET_c was generated. For the ET_c, it ranged from 3.2 to 6.0 mm/day

8.3.3 Performance of the Individual Tree-Level ET Estimation

As mentioned earlier, there were 80 sampling trees in the pomegranate field. Twenty of them were irrigated with 100% of ET_c, the same with the lysimeter tree.

Key Observation To validate the linear regression model on individual tree-level ET estimation, the authors compared the 20 sampling trees with the lysimeter daily ET_c for the UAV flight dates (almost the whole growing season). The trends of the daily ET_c for the 20 sampling trees (100% irrigation) and the lysimeter tree are shown in Fig. 8.4. Each "Serie" meant an individual tree in the field. Then, the boxplot of Fig. 8.5 was generated for analysis. Compared with the lysimeter tree, the linear regression model estimated tree-level ET with an R^2 and mean absolute error (MAE) of 0.9143 and 0.39 mm/day, respectively.

8.4 Conclusion

In this chapter, UAV flight missions were conducted to collect high-resolution multispectral imagery in a pomegranate orchard at USDA. Using the NDVI derived from the multispectral imagery, the authors applied a regression model between

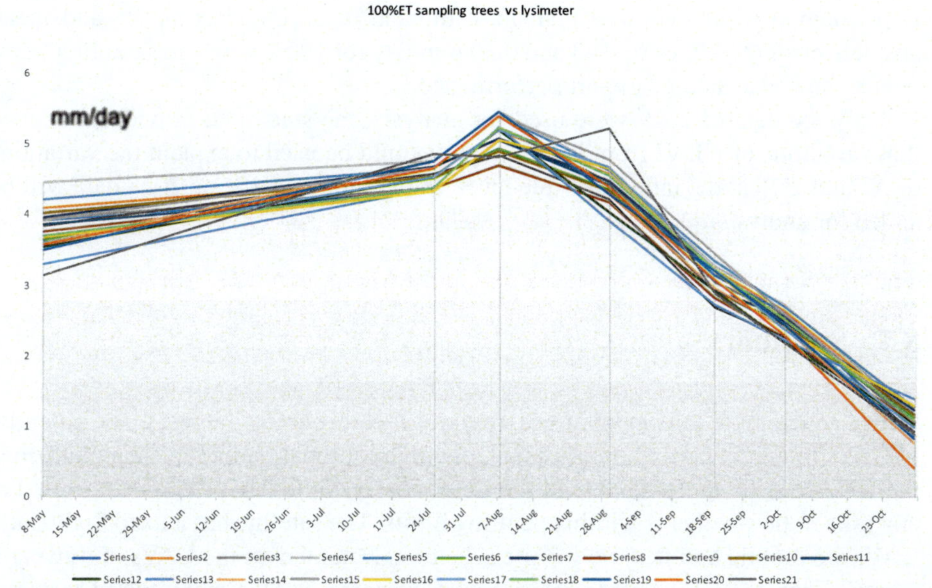

Fig. 8.4 100% ET irrigation sampling trees vs lysimeter tree. To validate the linear regression model on individual tree-level ET estimation, the authors compared the 20 sampling trees with the lysimeter daily ET_c for the UAV flight dates (almost the whole growing season). The trends of the daily ET_c for the 20 sampling trees (100% irrigation) and the lysimeter tree were shown. Each "Serie" meant an individual tree in the field

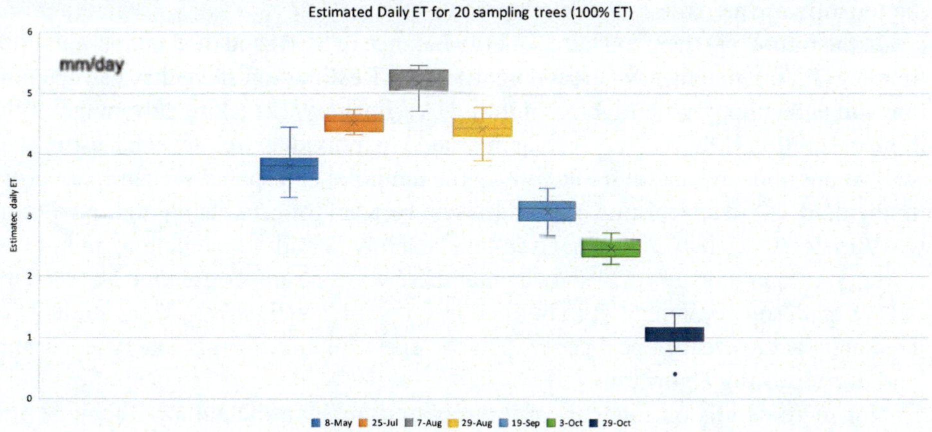

Fig. 8.5 The boxplot of 100% ET sampling trees vs lysimeter tree. Compared with the lysimeter tree, the linear regression model estimated tree-level ET with an R^2 and mean absolute error (MAE) of 0.9143 and 0.39 mm/day, respectively

NDVI and K_c to estimate the individual tree-level ET estimation. The linear regression model was $K_c(NDVI) = 1.6493NDVI - 0.6114$, which had an R^2 of 0.8865. Then, ET_c for all 100% ET irrigation trees was generated individually.

Experimental results showed that the estimated daily ET_c has an R^2 and mean absolute error (MAE) of 0.9143 and 0.39 mm/day for 100% irrigated sampling trees, which showed a state-of-the-art performance.

Only the data of 2019 were used for analysis; the study had provided evidence that variations of NDVI from UAV imagery could be used to explain the variations of K_c and ET_c at the individual tree level. In the future, 2017 and 2018 data will be added for analysis.

8.5 Summary

With a new remote sensing platform, researchers are gaining interest in the potential of UAVs in digital agriculture. Compared with traditional remote sensing platforms, the UAVs can be more flexible in the field. For example, UAVs can be operated at any time if the weather is within operating limits. The satellite has a fixed flight path, UAVs are mobile and flexible for site selection. Mounted on the UAVs, lightweight sensors, such as RGB cameras, multispectral cameras, and thermal infrared cameras, can be used to collect high-resolution images. Although there are many advantages to using UAVs, there are still challenges for UAVs when used to estimate ET. Many researchers fly the UAVs at different height, using specialized equipment and relying on data analysis expertise. As researchers try to understand and realize the potential of UAVs for ET estimation, efficient workflow, image processing, and better software are still under development.

In the future, (1) the two-source energy balance (TSEB) and dual temperature difference (DTD) models have great potential for ET estimation since they can separate the soil and canopy with high-resolution UAV imagery, (2) taking advantage of the high-resolution UAV imagery, research related to individual tree-level ET estimation will be possible and useful for analyzing the temporal and spatial variability of crops in the field, (3) deep learning algorithms can be used for processing high-resolution UAV imagery, such as individual tree-level canopy or soil segmentation, and (4) the authors' research results [32] showed that there was a strong correlation between the NDVI and crop coefficient at individual tree-level ET estimation. More studies can be conducted to create a new generation of vegetation index using machine learning and deep learning algorithms.

For more details on this part of the research, refer to the authors' book https://link.springer.com/book/10.1007/978-3-031-14937-5 [30].

References

1. Allen, R., Irmak, A., Trezza, R., Hendrickx, J.M., Bastiaanssen, W., Kjaersgaard, J.: Satellite-based ET estimation in agriculture using SEBAL and METRIC. Hydrol. Process. **25**(26), 4011–4027 (2011)

2. Allen, R.G., Pereira, L.S., Raes, D., Smith, M.: FAO Irrigation and Drainage Paper No. 56. Rome Food Agricult. Organiz. UN **56**(97), e156 (1998)
3. Allen, R.G., Tasumi, M., Morse, A., Trezza, R.: A Landsat-based energy balance and evapotranspiration model in western US water rights regulation and planning. Irrig. Drain. Syst. **19**(3–4), 251–268 (2005)
4. Allen, R.G., Tasumi, M., Trezza, R.: Satellite-based energy balance for mapping evapotranspiration with internalized calibration (METRIC)—Model. J. Irrig. Drain. Eng. **133**(4), 380–394 (2007)
5. Angus, D., Watts, P.: Evapotranspiration-How good is the Bowen ratio method? In: Developments in Agricultural and Managed Forest Ecology, vol. 13, pp. 133–150. Elsevier (1984)
6. Bastiaanssen, W.G., Menenti, M., Feddes, R., Holtslag, A.: A remote sensing surface energy balance algorithm for land (SEBAL). 1. Formulation. J. Hydrol. **212**, 198–212 (1998)
7. Brenner, C., Thiem, C.E., Wizemann, H.D., Bernhardt, M., Schulz, K.: Estimating spatially distributed turbulent heat fluxes from high-resolution thermal imagery acquired with a UAV system. Int. J. Remote Sens. **38**(8–10), 3003–3026 (2017)
8. Chávez, J.L., Gowda, P., Howell, T., Neale, C., Copeland, K.: Estimating hourly crop ET using a two-source energy balance model and multispectral airborne imagery. Irrig. Sci. **28**(1), 79–91 (2009)
9. Díaz-Varela, R., de la Rosa, R., León, L., Zarco-Tejada, P.: High-resolution airborne UAV imagery to assess olive tree crown parameters using 3D photo reconstruction: Application in breeding trials. Remote Sens. **7**(4), 4213–4232 (2015)
10. dos Reis, T., Monteiro, R., Garcia, E., Albuquerque, M., Espinoza, J., Ferreira, J.: Actual evapotranspiration estimated by orbital sensors, UAV and meteorological station for vineyards in the southern Brazil. In: IV Inovagri International Meeting (2017)
11. Elarab, M.: The application of unmanned aerial vehicle to precision agriculture: Chlorophyll, nitrogen, and evapotranspiration estimation. Ph.D. thesis, Utah State University (2016)
12. Ershadi, A., McCabe, M.F., Evans, J.P., Walker, J.P.: Effects of spatial aggregation on the multi-scale estimation of evapotranspiration. Remote Sens. Environ. **131**, 51–62 (2013)
13. Fritschen, L.J.: Accuracy of evapotranspiration determinations by the Bowen ratio method. Hydrol. Sci. J. **10**(2), 38–48 (1965)
14. Gonzalez-Dugo, V., Goldhamer, D., Zarco-Tejada, P.J., Fereres, E.: Improving the precision of irrigation in a pistachio farm using an unmanned airborne thermal system. Irrig. Sci. **33**(1), 43–52 (2015)
15. Hoffmann, H., Nieto, H., Jensen, R., Guzinski, R., Zarco-Tejada, P., Friborg, T.: Estimating evaporation with thermal UAV data and two-source energy balance models. Hydrol. Earth Syst. Sci. **20**(2), 697–713 (2016)
16. Hunsaker, D.J., Pinter, P.J., Barnes, E.M., Kimball, B.A.: Estimating cotton evapotranspiration crop coefficients with a multispectral vegetation index. Irrig. Sci. **22**(2), 95–104 (2003)
17. Kamble, B., Kilic, A., Hubbard, K.: Estimating crop coefficients using remote sensing-based vegetation index. Remote Sens. **5**(4), 1588–1602 (2013)
18. Kaplan, S., Myint, S.W., Fan, C., Brazel, A.J.: Quantifying outdoor water consumption of urban land use/land cover: Sensitivity to drought. Environ. Manag. **53**(4), 855–864 (2014)
19. Kustas, W., Anderson, M.: Advances in thermal infrared remote sensing for land surface modeling. Agricult. Forest Meteorol. **149**(12), 2071–2081 (2009)
20. Kustas, W., Norman, J.: Use of remote sensing for evapotranspiration monitoring over land surfaces. Hydrol. Sci. J. **41**(4), 495–516 (1996)
21. Kustas, W.P., Norman, J.M.: A two-source approach for estimating turbulent fluxes using multiple angle thermal infrared observations. Water Resour. Res. **33**(6), 1495–1508 (1997)
22. Liou, Y.A., Kar, S.: Evapotranspiration estimation with remote sensing and various surface energy balance algorithms—A review. Energies **7**(5), 2821–2849 (2014)
23. McShane, R.R., Driscoll, K.P., Sando, R.: A review of surface energy balance models for estimating actual evapotranspiration with remote sensing at high spatio temporal resolution over large extents. Scientific Investigations Report 2017–5087. Reston, VA: US Geological Survey. 19 p. (2017)

24. Monteiro, C.: Evapotranspiration estimate using energy balance two source model with UAV images: A study in vineyard. Am. J. Eng. Res. (AJER) **8**(21), 60–69 (2019)
25. Montibeller, Á.G.: Estimating energy fluxes and evapotranspiration of corn and soybean with an unmanned aircraft system in Ames, Iowa. Master's thesis, University of Northern Iowa (2017)
26. Moran, M.S., Jackson, R.D.: Assessing the spatial distribution of evapotranspiration using remotely sensed inputs. J. Environ. Qual. **20**(4), 725–737 (1991)
27. Nagler, P.L., Scott, R.L., Westenburg, C., Cleverly, J.R., Glenn, E.P., Huete, A.R.: Evapotranspiration on western US rivers estimated using the enhanced vegetation index from MODIS and data from Eddy covariance and Bowen ratio flux towers. Remote Sens. Environ. **97**(3), 337–351 (2005)
28. Nieto, H., Bellvert, J., Kustas, W.P., Alfieri, J.G., Gao, F., Prueger, J., Torres-Rua, A., Hipps, L.E., Elarab, M., Song, L.: Unmanned airborne thermal and multispectral imagery for estimating evapotranspiration in irrigated vineyards. In: Proceedings of the 2017 IEEE International Geoscience and Remote Sensing Symposium (IGARSS), pp. 5510–5513. IEEE (2017)
29. Nieto, H., Kustas, W.P., Torres-Rúa, A., Alfieri, J.G., Gao, F., Anderson, M.C., White, W.A., Song, L., del Mar Alsina, M., Prueger, J.H., et al.: Evaluation of TSEB turbulent fluxes using different methods for the retrieval of soil and canopy component temperatures from UAV thermal and multispectral imagery. Irrig. Sci. **37**(3), 389–406 (2019)
30. Niu, H., Chen, Y.: Towards Tree-level Evapotranspiration Estimation with Small UAVs in Precision Agriculture. Springer Nature, New York (2022)
31. Niu, H., Hollenbeck, D., Zhao, T., Wang, D., Chen, Y.: Evapotranspiration estimation with small UAVs in precision agriculture. Sensors **20**(22), 6427 (2020)
32. Niu, H., Wang, D., Chen, Y.: Estimating actual crop evapotranspiration using deep stochastic configuration networks model and UAV-based crop coefficients in a pomegranate orchard. In: Autonomous Air and Ground Sensing Systems for Agricultural Optimization and Phenotyping V. International Society for Optics and Photonics (2020)
33. Niu, H., Wang, D., Chen, Y.: Estimating crop coefficients using linear and deep stochastic configuration networks models and UAV-based normalized difference vegetation index (NDVI). In: Proceedings of the 2020 International Conference on Unmanned Aircraft Systems (ICUAS), pp. 1485–1490. IEEE (2020)
34. Niu, H., Zhao, T., Wang, D., Chen, Y.: Estimating evapotranspiration with UAVs in agriculture: A review. In: Proceedings of the ASABE Annual International Meeting. American Society of Agricultural and Biological Engineers (2019)
35. Niu, H., Zhao, T., Wang, D., Chen, Y.: A UAV resolution and waveband aware path planning for onion irrigation treatments inference. In: 2019 International Conference on Unmanned Aircraft Systems (ICUAS), pp. 808–812. IEEE (2019)
36. Norman, J.M., Kustas, W.P., Humes, K.S.: Source approach for estimating soil and vegetation energy fluxes in observations of directional radiometric surface temperature. Agricult. Forest Meteorol. **77**(3–4), 263–293 (1995)
37. Ortega-Farías, S., Ortega-Salazar, S., Poblete, T., Kilic, A., Allen, R., Poblete-Echeverría, C., Ahumada-Orellana, L., Zuñiga, M., Sepúlveda, D.: Estimation of energy balance components over a drip-irrigated olive orchard using thermal and multispectral cameras placed on a helicopter-based unmanned aerial vehicle (UAV). Remote Sens. **8**(8), 638 (2016)
38. Park, S.: Estimating plant water stress and evapotranspiration using very-high-resolution (VHR) UAV imagery. Ph.D. thesis, University of Melbourne (2018)
39. Phene, C., McCormick, R., Davis, K., Pierro, J., Meek, D.: A lysimeter feedback system for precise evapotranspiration measurement and irrigation control. Trans. ASAE **32**(2), 477–484 (1989)
40. Quattrochi, D.A., Luvall, J.C.: Thermal infrared remote sensing for analysis of landscape ecological processes: Methods and applications. Landscape Ecol. **14**(6), 577–598 (1999)
41. Reyes-Gonzalez, A., Hay, C., Kjaersgaard, J., Neale, C.: Use of remote sensing to generate crop coefficient and estimate actual crop evapotranspiration. In: Proceedings of the 2015 ASABE

Annual International Meeting, p. 1. American Society of Agricultural and Biological Engineers (2015)
42. Su, Z.: The surface energy balance system (SEBS) for estimation of turbulent heat fluxes. Hydrol. Earth Syst. Sci. **6**(1), 85–100 (2002)
43. Swain, K.C., Thomson, S.J., Jayasuriya, H.P.: Adoption of an unmanned helicopter for low-altitude remote sensing to estimate yield and total biomass of a rice crop. Trans. ASABE **53**(1), 21–27 (2010)
44. Trout, T.J., Johnson, L.F., Gartung, J.: Remote sensing of canopy cover in horticultural crops. HortScience **43**(2), 333–337 (2008)
45. Verstraeten, W., Veroustraete, F., Feyen, J.: Assessment of evapotranspiration and soil moisture content across different scales of observation. Sensors **8**(1), 70–117 (2008)
46. Wang, D., Ayars, J., Tirado-Corbala, R., Makus, D., Phene, C., Phene, R.: Water and nitrogen management of young and maturing pomegranate trees. In: III International Symposium on Pomegranate and Minor Mediterranean Fruits 1089, pp. 395–401 (2013)
47. Wetzel, P.J., Chang, J.T.: Evapotranspiration from nonuniform surfaces: A first approach for short-term numerical weather prediction. Month. Weather Rev. **116**(3), 600–621 (1988)
48. Wu, J., Wang, D.: Estimating evaporation coefficient during two-stage evaporation from soil surfaces. Soil Sci. **170**(4), 235–243 (2005)
49. Xia, T., Kustas, W.P., Anderson, M.C., Alfieri, J.G., Gao, F., McKee, L., Prueger, J.H., Geli, H.M., Neale, C.M., Sanchez, L., et al.: Mapping evapotranspiration with high-resolution aircraft imagery over vineyards using one-and two-source modeling schemes. Hydrol. Earth Syst. Sci. **20**(4), 1523 (2016)
50. Xu, C.Y., Singh, V.: Evaluation of three complementary relationship evapotranspiration models by water balance approach to estimate actual regional evapotranspiration in different climatic regions. J. Hydrol. **308**(1–4), 105–121 (2005)
51. Zarco-Tejada, P.J., González-Dugo, V., Williams, L., Suárez, L., Berni, J.A., Goldhamer, D., Fereres, E.: A PRI-based water stress index combining structural and chlorophyll effects: Assessment using diurnal narrow-band airborne imagery and the CWSI thermal index. Remote Sens. Environ. **138**, 38–50 (2013)
52. Zhang, H., Anderson, R.G., Wang, D.: Satellite-based crop coefficient and regional water use estimates for Hawaiian sugarcane. Field Crops Res. **180**, 143–154 (2015)
53. Zhang, H., Wang, D., Ayars, J.E., Phene, C.J.: Biophysical response of young pomegranate trees to surface and sub-surface drip irrigation and deficit irrigation. Irrig. Sci. **35**(5), 425–435 (2017)
54. Zhao, T., Chen, Y., Ray, A., Doll, D.: Quantifying almond water stress using unmanned aerial vehicles (UAVs): Correlation of stem water potential and higher order moments of non-normalized canopy distribution. In: ASME 2017 International Design Engineering Technical Conferences and Computers and Information in Engineering Conference. American Society of Mechanical Engineers (2017)
55. Zhao, T., Doll, D., Chen, Y.: Better almond water stress monitoring using fractional-order moments of non-normalized difference vegetation index. In: 2017 ASABE Annual International Meeting, p. 1. American Society of Agricultural and Biological Engineers (2017)
56. Zhao, T., Niu, H., de la Rosa, E., Doll, D., Wang, D., Chen, Y.: Tree canopy differentiation using instance-aware semantic segmentation. In: 2018 ASABE Annual International Meeting. American Society of Agricultural and Biological Engineers (2018)
57. Zhao, T., Stark, B., Chen, Y., Ray, A.L., Doll, D.: A detailed field study of direct correlations between ground truth crop water stress and normalized difference vegetation index (NDVI) from small unmanned aerial system (sUAS). In: 2015 International Conference on Unmanned Aircraft Systems (ICUAS), pp. 520–525. IEEE (2015)
58. Zhao, T., Wang, Z., Yang, Q., Chen, Y.: Melon yield prediction using small unmanned aerial vehicles. In: Autonomous Air and Ground Sensing Systems for Agricultural Optimization and Phenotyping II, vol. 10218, p. 1021808. International Society for Optics and Photonics (2017)

Chapter 9
Individual Tree-Level Water Status Inference Using High-Resolution UAV Thermal Imagery and Complexity-Informed Machine Learning

Abstract This chapter introduces an innovative methodology for inferring the water status of individual trees through the integration of high-resolution Unmanned Aerial Vehicle (UAV) thermal imagery and Complexity-informed Machine Learning (CIML). The chapter opens with an insightful introduction, emphasizing the critical role of individual tree-level water status assessment for digital agriculture. Materials and methods are meticulously described, covering the experimental site, irrigation management practices, ground truth data collection through infrared canopy and air temperature measurements, acquisition of thermal infrared remote sensing data using UAVs, and the application of Complexity-Informed Machine Learning (CIML). The principle of tail matching is introduced as a key component of the methodology, along with various machine learning classification algorithms employed for analysis. Results and discussion present a comprehensive analysis, including a comparison of canopy temperature per tree based on ground truth and UAV thermal imagery, the relationship between temperature differentials and irrigation treatment, and the classification performance of CIML on different irrigation treatment levels. These findings provide valuable insights into the efficacy of the proposed methodology in accurately assessing the water status of individual trees. The chapter concludes with a succinct summary, encapsulating key findings and their implications.

9.1 Introduction

There is around 11,000 ha of pomegranate in California due to its resistance to drought and high economic value [28]. Considering the recurring water shortage in California, it is essential to find effective methods to optimize irrigation water use. Research results suggested that the evapotranspiration (ET) estimation is one of the most critical factors to help manage water use efficiency in agriculture [1]. The temporal and spatial mapping of ET can identify variations in the field, which helps to evaluate the status of crop water [14].

The temperature of the tree canopy from the infrared thermometer (IRT) sensors is an effective tool to detect plant water stress. Research has been conducted on

H. Niu, Y. Chen, *Smart Big Data in Digital Agriculture Applications*, Agriculture Automation and Control, https://doi.org/10.1007/978-3-031-52645-9_9

crops and trees to relate the ΔT to irrigation management. The main reason is that a significant increase in the midday infrared canopy to air temperature difference (Δ T) will indicate stomatal closure and water stress conditions [11–13]. For example, Zhang et al. evaluated the performance of using Δ T to manage postharvest deficit irrigation of nectarine trees [27]. The results demonstrated that the measured ΔT values above the tree canopy showed consistent differences between the levels of irrigation treatment. Clawson et al. used the variability of the canopy temperature and the average canopy temperature to schedule irrigation in corn (*Zea mays* L.). They remarked that canopy temperature variability could show the plant water stress and the need to schedule an irrigation event [6]. Furthermore, Wang et al. investigated the infrared canopy temperature of early ripening peach trees under postharvest deficit irrigation and monitored the stem water potential. The strong correlation between the stem water potential and ΔT ($R^2 \approx 0.7$) indicated that the canopy temperature could be used for the estimation of water stress.

However, little research could be found in the literature on using midday ΔT derived from UAV thermal infrared (TIR) image as a primary input for mapping irrigation treatment levels of a pomegranate field at individual tree levels. This article evaluated the feasibility and performance of using midday ΔT (UAV-TIR) and machine learning algorithms to spatialize irrigation treatments. Recently, UAVs have emerged as powerful platforms in agricultural applications, such as irrigation management [20, 30], and water stress estimation [29]. With lightweight sensors mounted on UAVs, high spatial and temporal resolution images have been taken in massive amounts with low cost [15, 31]. Due to its lightweight and low power consumption, the thermal camera has been commonly used in agricultural research [10, 23]. The spatial resolution of the UAV-based thermal imagery can be at the centimeter level and help identify, standardize, and validate methods to calculate the spatial variability for clumped canopy structures, such as trees and vines [21].

Machine learning (ML) models have been widely used in real-world applications, for example, image processing [32], natural language processing [9], and digital agriculture [19]. ML algorithms can simplify a solution and perform better than traditional statistical approaches, which may require more rules for hand-tuning. However, training ML models may require a large amount of data, which may not always be available for scientific problems [22]. Then, the smaller dataset may cause ML models to lack robustness and cannot guarantee convergence. Therefore, in this chapter, the authors proposed the concept of complexity-informed machine learning (CIML) and the principle of tail matching (POTM). The original dataset can exhibit a heavy-tailed distribution phenomenon, and the tail-index analysis can be used for ML algorithms [18, 26]. Specifically, tail information in the variability and diversity of the training dataset should indicate the representativeness of the data. In this sense, we can expect a "smaller dataset" rather than "big data" for ML under the same performance requirement. In summary, we pursue "tail matching" between the dataset and the ML algorithms.

The **objectives of this study** were (1) evaluated the reliability of the UAV thermal camera on individual tree canopy temperature measurements, (2) investigated and validated the approaches of irrigation treatment inference using UAV-based ΔT at individual tree level, and (3) demonstrated the performance of the CIML models in inference of irrigation treatment. The **significant contributions of this chapter** were (1) developed a reliable tree-level water stress detection method using UAV-based high-resolution thermal images and (2) proposed the concept of CIML and proved its performance on the classification of tree-level irrigation treatments. The rest of the chapter is organized as follows. Section 9.2 introduces the materials and methods being used for UAV-based irrigation treatment inference. The results and discussion are presented in Sect. 9.3. In Sect. 9.4, the authors draw conclusive remarks.

9.2 Materials and Methods

9.2.1 Experimental Site and Irrigation Management

The study was carried out in a 1.3 ha pomegranate field in 2019 at the USDA-ARS San Joaquin Valley Agricultural Sciences Center in Parlier, CA (36.594°N, 119.512°W). The pomegranate (*Punica granatum* L., cv 'Wonderful') was planted in 2010 with a 5 m spacing between rows and a 2.75 m within-row tree spacing [28]. The soil type was a Hanford fine sandy loam (coarse-loamy, mixed, and thermic Typic Xerorthents). There are also two large weighing lysimeters, which are 2 m×4 m by 3 m in depth and have a resolution of 0.1 mm of water loss. As shown in Fig. 9.1, the weighing lysimeters are located in the center of the pomegranate field. The experimental site was randomly designed into 16 blocks, with four replications, to test four irrigation rates on the pomegranate growth. As measured by the lysimeter, the irrigation volumes were set up as 35%, 50%, 75%, and 100% of ET_c. The trees in the lysimeter were irrigated at the 100% treatment

Fig. 9.1 The pomegranate field at the USDA-ARS (36.594°N, 119.512°W). The weighing lysimeter is located in the center of the pomegranate field, marked as a red box. The blue marks are where the 14 IRT sensors were installed

Fig. 9.2 The IRT sensor was installed 4.5 m above the soil surface, with an FOV of 20°. A quadcopter was used as the low-cost UAV platform (less than $1000) to collect high-resolution thermal images at a height of 60 m

level. For each irrigation treatment block, there were three rows with around 15 trees per row. Only the central row of each block was used as the experimental row. The height of trees was pruned and maintained at approximately 3 m.

9.2.2 Ground Truth: Infrared Canopy and Air Temperature

The temperature of the tree canopy was measured with 14 IRT sensors (Model SI-100 series, Apogee Instruments, Inc., Logan, UT), which were installed 4.5 m above the soil surface. The field of view (FOV) of the IRT sensor was 20° (Fig. 9.2). The air temperature and relative humidity were also measured with a sensor at the experimental site. Then the authors could evaluate the performance of using the midday infrared canopy to air temperature difference (ΔT) to detect or classify the deficit irrigation of pomegranate trees.

9.2.3 Thermal Infrared Remote Sensing Data

A quadcopter named "Foxtech Hover" was used as a low-cost UAV platform (less than $1000) to collect high-resolution thermal images at the height of 60 m. The

UAV was equipped with a highly efficient power system, including a T-Motor MN3508 KV380 motor, 1552 folding propellers, and Foxtech Multi-Pal 40A OPTP ESC, to ensure long flight time. The UAV included a Pixhawk flight controller, GPS, and telemetry antennas, enabling it to fly over the pomegranate field by waypoints mode (designed using Mission Planner software). The lithium polymer battery of Hover had a capacity of 9500 mAh, which could support our 30-minute flight mission with the thermal camera mounted on the UAV. The thermal camera ICI 9640 P (Infrared Cameras Inc, Beaumont, TX, USA)[1] was equipped with a UAV to collect thermal images for the experimental field. The sensor has a resolution of 640×480 pixels. The spectral band ranges from 7 μm to 14 μm. The dimension of the thermal camera is 34 mm×30 mm×34 mm. The accuracy is designed to be ± 2 °C. A Raspberry Pi Model B computer (Raspberry Pi Foundation, Cambridge, UK) was used to trigger the thermal camera during UAV flight missions.

9.2.3.1 UAV Thermal Image Collection and Processing

The authors used Mission Planner to program all flight missions. The flight height was set at 60 m. The overlapping of UAV imagery was set up as 80% so that the UAV imagery of the pomegranate could be stitched together more successfully during image processing. The UAV was flying at noon in clear sky conditions to minimize the shading effect on the thermal images. Since the thermal camera type is uncooled, it usually takes around 20 minutes to warm up the thermal camera before flight missions. To calibrate the thermal camera, the authors took thermal images of ice water immediately before and after the flight missions as the reference temperature. After the flight missions, all UAV thermal images were stitched together to generate the orthomosaic images in Metashape (Agisoft LLC, Russia). Preselection is recommended because it can speed up the processing of large datasets. Building a dense cloud can reconstruct a more accurate surface, improving the quality of the final orthomosaic. Higher quality usually results in a more accurate surface, which means a more significant number of points. Building a digital elevation model (DEM) allows generating an accurate surface, which can be used as a source for orthomosaic generation. The above steps will shorten the data processing time compared to the Build Mesh operation, because Build Mesh is usually used for a more complex surface.

[1] Mention of trade names or commercial products in this publication is solely to provide specific information and does not imply recommendation or endorsement by the University of California or the US Department of Agriculture. The University of California and the USDA are equal opportunity providers and employers.

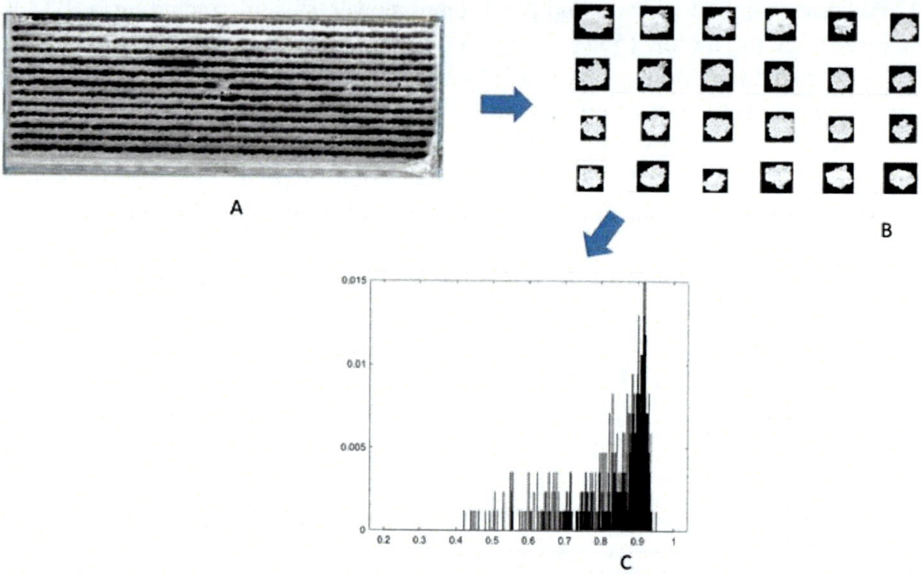

Fig. 9.3 **(a)** All of the UAV thermal images were stitched together to generate the orthomosaick images in Metashape (Agisoft LLC, Russia). **(b)** To obtain the individual tree canopy images of the 250 sampling trees, the authors used the SVM for classifying the tree canopy. **(c)** Histogram was generated for each tree canopy to check the variability of each tree canopy temperature

9.2.3.2 Tree Canopy Segmentation Using Support Vector Machine (SVM)

There were 746 trees in total for the pomegranate field. As mentioned above, there were three rows with around 15 trees per row for each irrigation treatment block. Only the central row of each block was used as the experimental row. To obtain individual tree canopy images of the 250 sampling trees (Fig. 9.3b), the authors used SVM to classify the tree canopy with ArcGIS Pro. Using the SVM classifier could map the input data vectors into a higher dimensional feature space. Then, the SVM optimally separated the data into different classes. Since the file size of UAV-based thermal imagery was large, the SVM classifier was adopted, which was less susceptible to noise, unbalanced numbers, or the size of training sites within each class. All the sampling trees were successfully segmented using the SVM classifier. Then the mean, variance, and histogram information (Fig. 9.3c) were calculated using MATLAB 2021b as input features for the CIML models.

9.2.4 Complexity-Informed Machine Learning (CIML)

When the authors discuss complexity-informed machine learning, they focus on variability analysis using the histogram information of individual tree canopy.

Variability refers to several properties of the ML dataset. First, it refers to the number of inconsistencies in the data which need to be understood by using anomaly- and outlier-detection methods for any meaningful analysis to be performed. Second, variability can also refer to diversity. For example, when the authors studied the individual tree-level ET estimation [21], it turned out that the evapotranspiration for each tree is very close to each other. The reason was that the mean pixel values were used for data analysis, making the ET classification challenging to implement. Considering the complexity of each tree canopy, embedding complex information into the ML training process may have great potential to detect or classify pomegranate tree deficit irrigation. To analyze the complex information in each tree, we need to use tail-index analysis methods.

9.2.5 Principle of Tail Matching

In the probability theory, heavy-tailed distributions are PDFs whose tails do not decay exponentially [2]. The distribution of a random variable with real values X is said to have a heavy right tail if the tail probabilities P decay more slowly than those of any exponential distribution. Consequently, they have more weight in their tails than does an exponential distribution

$$\lim_{x \to \infty} \left(\frac{P(X > x)}{e^{-\lambda x}} \right) = \infty, \tag{9.1}$$

for every $\lambda > 0$ [24]. The tail information in the training dataset variability and diversity should be used to indicate the data reprcsentativeness. In this chapter, the Generalized Pareto (GP) distribution was developed to model tail index for individual tree canopy thermal imagery.

9.2.5.1 Pareto Distribution

A random variable is said to have a Pareto probability density function (PDF) if its cumulative distribution function (CDF) is

$$F(x) = \begin{cases} 1 - (\frac{b}{x})^a, x \geq b, \\ 0, x < b, \end{cases} \tag{9.2}$$

where $b > 0$ is the scale parameter, and $a > 0$ is the shape parameter which is the Pareto's index of inequality [7]. The tail data of the tree canopy temperature were fitted using the generalized Pareto (GP) distribution by the maximum likelihood estimation. Many fitting models may agree well with the data in high-density regions but poorly in low-density areas. However, in many applications, fitting the data in the tail may also contribute to model performance. The GP was developed as a distribution that can model tails of a wide variety of distributions based on theoretical arguments.

9.2.6 Machine Learning Classification Algorithms

Several classification algorithms were adopted to evaluate the detection performance for irrigation treatment levels. "Neural Net," "Support Vector Machine (SVM)," "Random Forest," "AdaBoost," "Nearest Neighbors," "Gaussian Process," "Naive Bayes," "Quadratic Discriminant Analysis," and "Decision Tree" were chosen as the classification algorithms. Some of them were briefly introduced for reference.

In the "Neural Net" library, a multi-layer perceptron (MLP) classifier was used. This model optimized the log-loss function using stochastic gradient descent. MLP trained iteratively because the partial derivatives of the loss function concerning the model parameters were computed to update the parameters at every step [25]. The SVMs are a set of supervised learning methods used for classification, regression, and outlier detection. The SVMs are effective in high-dimensional spaces and effective in cases where the number of dimensions is greater than the number of samples [5]. For the "Random Forest" classifier, it is a meta-estimator that fits several decision tree classifiers on various sub-samples of the dataset and adopts averaging to improve the predictive accuracy and control overfitting [4].

An "AdaBoost classifier" is also a meta-estimator that begins by fitting a classifier on the original dataset. Then, the model fits additional copies of the classifier on the same dataset but where the weights of incorrectly classified instances are adjusted such that subsequent classifiers focus more on complex cases [16]. The "Nearest Neighbors" method is to figure out a predefined number of training samples closest in the distance to the new point and predict the label from those [8]. The samples can be a constant (k-nearest neighbor learning) or vary based on the local density of points (radius-based neighbor learning). Despite its simplicity, the nearest neighbors method has been successfully applied for many research problems, such as the handwritten digit classification. As a non-parametric method, it is often successful in classification situations where the decision boundary is very irregular.

The "Decision Trees" method is also a non-parametric supervised learning method commonly adopted for classification problems. The objective is to create a model that predicts the value of a target variable by learning simple decision rules inferred from the data features [17]. A tree can be seen as a piecewise constant approximation. "Decision Trees" usually use a white box model, which means the explanation for the condition is easily explained by Boolean logic if a given situation is observable in a model. In contrast, results may be more challenging to interpret for a black box model, such as an artificial neural network.

9.3 Results and Discussion

9.3.1 Comparison of Canopy Temperature Per Tree Based on Ground Truth and UAV Thermal Imagery

To evaluate the reliability of UAV thermal remote sensing, the authors first compared the canopy temperature per tree acquired by IRT sensors and the UAV thermal camera. The correlation between the canopy temperature per tree measured by the IRT sensors and the UAV thermal camera was shown by their scatter-related plot and the established regression equation (Fig. 9.4). The coefficient of determination (R^2) was 0.8668, which indicated that the difference between the ground truth and the UAV thermal camera was acceptable. The method was reliable for monitoring tree-level canopy temperature.

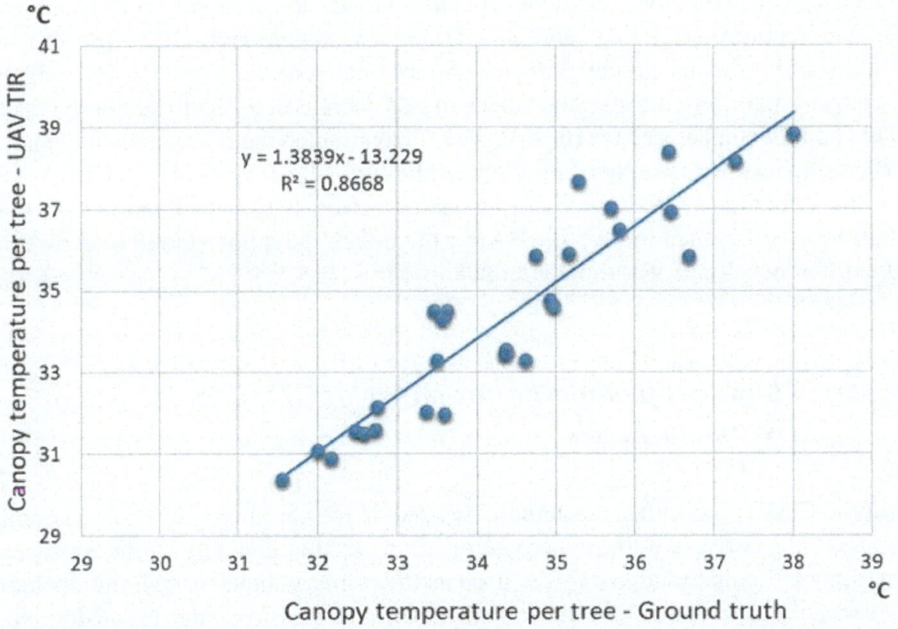

Fig. 9.4 The correlation between the canopy temperature per tree measured by the IRT sensors and UAV thermal camera. The coefficient of determination (R^2) was 0.8668, which indicated that the difference between the ground truth and the UAV thermal camera was acceptable. The method was reliable for monitoring tree-level canopy temperature

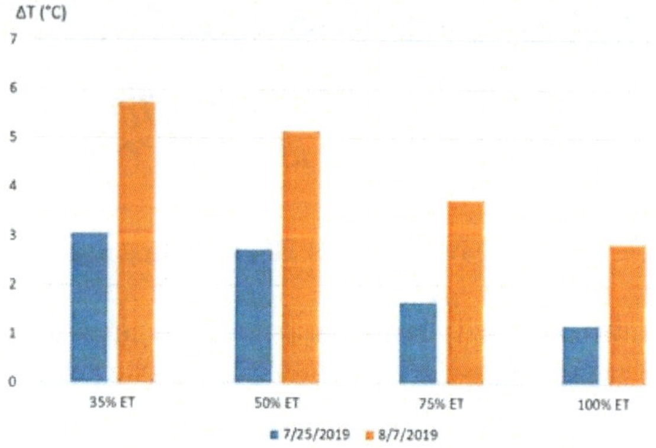

Fig. 9.5 The ΔT was significantly higher in the 35% irrigation treatment than the 100% irrigation treatment on different days. The values of ΔT decreased as the irrigation increased. This finding emphasized the importance of irrigation on the tree canopy temperature response

9.3.2 The Relationship Between ΔT and Irrigation Treatment

The effect of irrigation treatment on canopy to air temperature (ΔT) was plotted in this section (7-25-2019 and 8-7-2019). As shown in Fig. 9.5, the ΔT was significantly higher in the 35% irrigation treatment than the 100% irrigation treatment on different days. The values of ΔT decreased as the irrigation increased. This finding emphasized the importance of irrigation on the tree canopy temperature response. Several researchers reported similar results [3, 27, 28]. At the USDA-ARS, all the pomegranate trees were fully irrigated before 2012, which did not show any significant difference for ΔT [28]. After the deficit irrigation started in early 2012, the difference of ΔT was more significant.

9.3.3 The Classification Performance of CIML on Irrigation Treatment Levels

For the CIML algorithms, the authors focused on the variability analysis. Variability refers to the individual tree canopy temperature spatial diversity. Different types of tree canopy temperature data were used as the primary input for training, including (1) mean and variance, (2) tail index, mean, and variance, and (3) histogram of tree canopy temperature. The tree canopy temperature of 250 sampling trees was distributed as 75% for training and 25% for testing using the *train_test_split* method. For evaluating the trained CIML models, a confusion matrix was used to compare the performances of different classifiers. A confusion matrix was a summary of prediction results on a classification problem. The number of correct and incorrect predictions was tallied with count values and divided into classes. The confusion matrix provided insight into not only the errors being made by a classifier but, more importantly, the types of errors that were being made as well. "True label" meant the ground truth of ET_c-based irrigation treatment levels. "Predicted label" identified the irrigation treatment levels predicted by the trained model. To simplify the visualization, 30% and 50% ET irrigations were labeled as "0," denoting low-level irrigation, and 75% and 100% ET irrigations were labeled as "1," which meant high-level irrigation.

The trained models had distinct test performance for irrigation treatment prediction at tree level (Fig. 9.6, Table 9.1, and Fig. 9.7). First of all, the most important finding was that using the UAV-based tree canopy to air temperature (ΔT) and machine learning algorithms could successfully classify the irrigation treatment or water stress at the individual tree level. The research results demonstrated that ΔT was highly related to irrigation management. As mentioned earlier, the main reason was that a significant increase ΔT would indicate stomata closure and water stress conditions [11–13]. Thus, UAV-based thermal remote sensing is a reliable tool for tree irrigation management. The results were highly consistent for different methods. For example, when histogram information was used for training and testing. All the methods showed a state-of-the- art performance, with an overall accuracy of 87%. The "Naive Bayes" had the highest accuracy of 0.90.

Another finding was that tail-index information had great potential to benefit training and testing performance. The mean and variance were a simplification of complex information. By adding the tail information into the training dataset, the prediction accuracy of some methods was increased, as shown in Table 9.1. It inspired us that the tail information in the training dataset variability and diversity should be used to indicate the data representativeness. Then, with more complex information, the histogram information of tree canopy temperature had the best prediction accuracy, without a doubt. In summary, all three situations had the overall accuracy above 80%, mainly because the ΔT was very sensitive to irrigation treatments.

9.4 Summary

The aim of this chapter was for irrigation treatment levels inference in the pomegranate field at the individual tree level by using UAV-based thermal images and machine learning algorithms. The authors collected the ΔT by using a UAV-based high-resolution thermal camera. Then, CIML algorithms were adopted for the tree-level irrigation treatment classification problem. The authors developed a reliable tree-level irrigation treatment inference method using UAV-based high-resolution thermal images. The research results showed that the best classification accuracy of irrigation treatment levels was 90% when the "Naive Bayes" method was adopted. The results of this research supported the idea that a significant increase in the midday infrared canopy to air temperature difference (ΔT) will indicate stomata closure and water stress conditions. The authors also proposed the concept of CIML and proved its performance on the classification of tree-level irrigation treatments. CIML models have great potential for future agriculture research. With more complex information, it will benefit the training and testing process of machine learning algorithms.

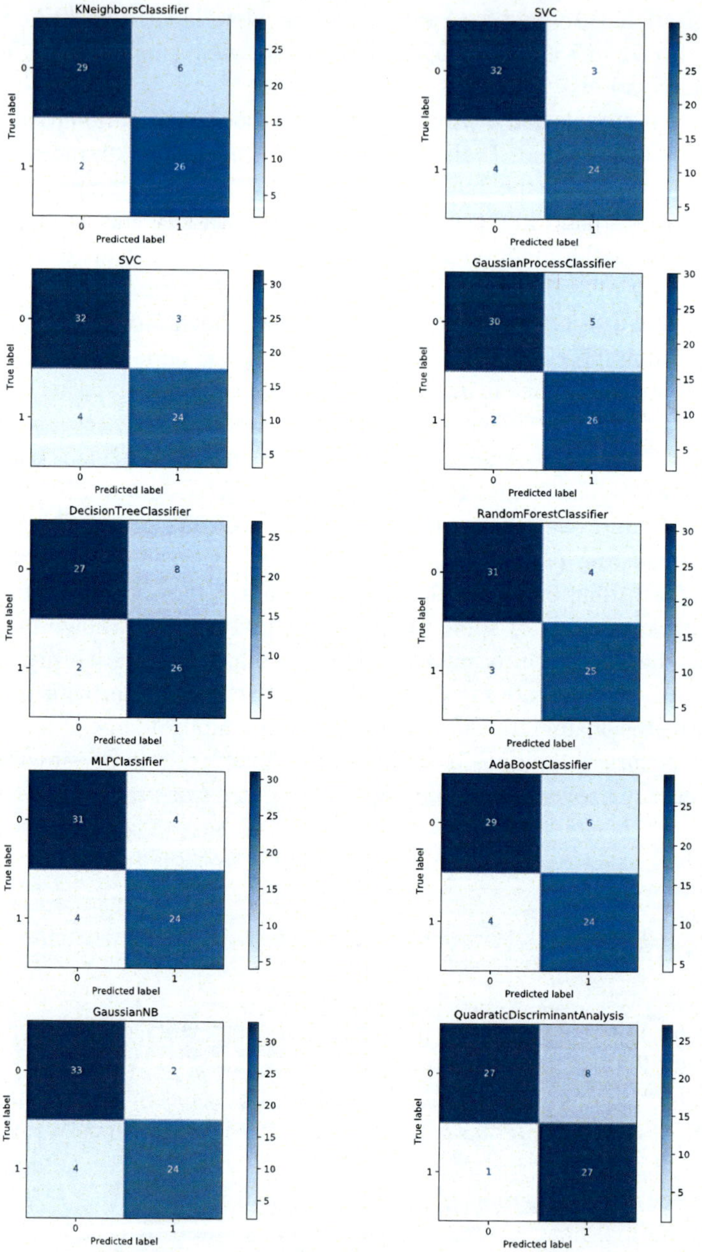

Fig. 9.6 The summary of prediction results using histogram information on the tree-level irrigation treatment inference. "True label" meant the ground truth of ET_c-based irrigation treatment levels. "Predicted label" identified the irrigation treatment levels predicted by the trained model. To simplify the visualization, 30% and 50% ET irrigations were labeled as "0," denoting low-level irrigation, and 75% and 100% ET irrigations were labeled as "1", which meant high-level irrigation

Table 9.1 The classification performance of CIML algorithms on irrigation treatment levels at individual tree level. All the methods showed a state-of-the-art performance, with an overall accuracy of 87%. The "Naive Bayes" had the highest accuracy of 0.90

Classification methods	Prediction accuracy (histogram)	Prediction accuracy (mean, variance, and tail index)	Prediction accuracy (mean + variance)
"KNeighborsClassifier"	0.87	0.86	0.84
"Linear SVM"	0.89	0.86	0.84
"RBF SVM"	0.89	0.84	0.84
"Gaussian Process"	0.89	0.86	0.86
"Decision Tree"	0.84	0.89	0.87
"Random Forest"	0.89	0.87	0.89
"Neural Net"	0.87	0.89	0.44
"AdaBoost"	0.84	0.87	0.89
"Naive Bayes"	**0.90**	0.81	0.68
"QDA"	0.86	0.83	0.73

The bold value is the model with best prediction performance.

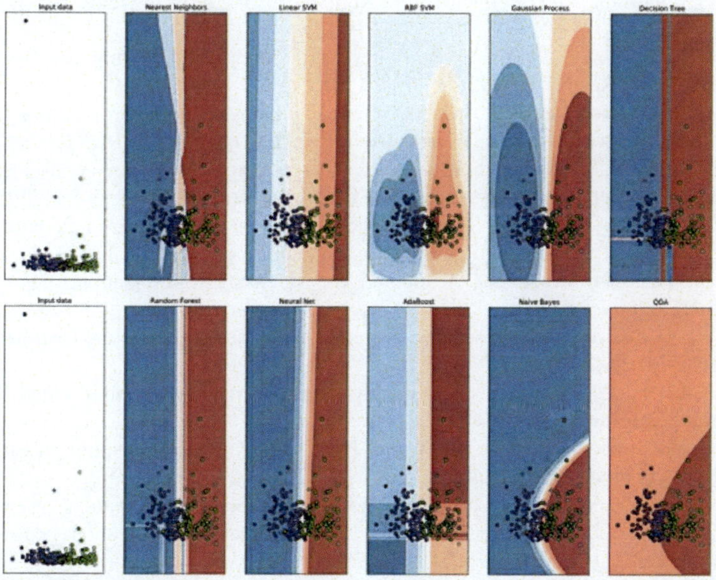

Fig. 9.7 The test performance for the histogram dataset. The t-distributed stochastic neighbor embedding (TSNE) method was used for data visualization, which learned the most critical axes between the classes. The axes were then used to define the hyperplane to project the high-dimensional training data into two dimensions, which gained important insight by visually detecting patterns. The x-axis and y-axis had no scale because of hyperplane projection. The irrigation treatment levels were successfully clustered into low level (blue) and high level (green)

References

1. Allen, R.G., Pereira, L.S., Raes, D., Smith, M.: FAO Irrigation and Drainage Paper No. 56. Rome Food Agricult. Organiz. UN **56**(97), e156 (1998)
2. Asmussen, S.: Steady-state properties of $GI/G/1$. Appl. Probab. Queues, 266–301 (2003)
3. Ballester, C., Castel, J., Jiménez-Bello, M., Castel, J., Intrigliolo, D.: Thermographic measurement of canopy temperature is a useful tool for predicting water deficit effects on fruit weight in citrus trees. Agricult. Water Manag. **122**, 1–6 (2013)
4. Breiman, L.: Random forests. Mach. Learn. **45**(1), 5–32 (2001)
5. Chang, C.C., Lin, C.J.: LIBSVM: A library for support vector machines. ACM Trans. Intell. Syst. Technol. (TIST) **2**(3), 1–27 (2011)
6. Clawson, K.L., Blad, B.L.: Infrared thermometry for scheduling irrigation of corn 1. Agron. J. **74**(2), 311–316 (1982)
7. Geerolf, F.: A theory of Pareto distributions. UCLA Manuscript (2016)
8. Goldberger, J., Hinton, G.E., Roweis, S., Salakhutdinov, R.R.: Neighbourhood components analysis. Adv. Neural Inf. Process. Syst. **17**, 513–520 (2004)
9. Hirschberg, J., Manning, C.D.: Advances in natural language processing. Science **349**(6245), 261–266 (2015)
10. Hoffmann, H., Nieto, H., Jensen, R., Guzinski, R., Zarco-Tejada, P., Friborg, T.: Estimating evaporation with thermal UAV data and two-source energy balance models. Hydrol. Earth Syst. Sci. **20**(2), 697–713 (2016)
11. Idso, S.B., Jackson, R.D., Reginato, R.J.: Remote-sensing of crop yields. Science **196**(4285), 19–25 (1977)
12. Jackson, R.D., Idso, S., Reginato, R., Pinter Jr, P.: Canopy temperature as a crop water stress indicator. Water Resour. Res. **17**(4), 1133–1138 (1981)
13. Jackson, R.D., Reginato, R., Idso, S.: Wheat canopy temperature: A practical tool for evaluating water requirements. Water Resour. Res. **13**(3), 651–656 (1977)
14. Kaplan, S., Myint, S.W., Fan, C., Brazel, A.J.: Quantifying outdoor water consumption of urban land use/land cover: Sensitivity to drought. Environ. Manag. **53**(4), 855–864 (2014)
15. Khanal, S., Fulton, J., Shearer, S.: An overview of current and potential applications of thermal remote sensing in precision agriculture. Comput. Electron. Agricult. **139**, 22–32 (2017)
16. Kingma, D.P., Ba, J.: Adam: A method for stochastic optimization. Preprint (2014). arXiv:1412.6980
17. Loh, W.Y.: Classification and regression trees. Wiley Interdiscip. Rev. Data Mining Knowl. Disc. **1**(1), 14–23 (2011)
18. Niu, H., Chen, Y., West, B.J.: Why do big data and machine learning entail the fractional dynamics? Entropy **23**(3), 297 (2021)
19. Niu, H., Hollenbeck, D., Zhao, T., Wang, D., Chen, Y.: Evapotranspiration estimation with small UAVs in precision agriculture. Sensors **20**(22), 6427 (2020)
20. Niu, H., Zhao, T., Wang, D., Chen, Y.: A UAV resolution and waveband aware path planning for onion irrigation treatments inference. In: 2019 International Conference on Unmanned Aircraft Systems (ICUAS), pp. 808–812. IEEE (2019)
21. Niu, H., Zhao, T., Wei, J., Wang, D., Chen, Y.: Reliable tree-level evapotranspiration estimation of pomegranate trees using lysimeter and UAV multispectral imagery. In: 2021 IEEE Conference on Technologies for Sustainability (SusTech), pp. 1–6. IEEE (2021)
22. Raissi, M., Perdikaris, P., Karniadakis, G.E.: Physics-informed neural networks: A deep learning framework for solving forward and inverse problems involving nonlinear partial differential equations. J. Comput. Phys. **378**, 686–707 (2019)
23. Ribeiro-Gomes, K., Hernández-López, D., Ortega, J.F., Ballesteros, R., Poblete, T., Moreno, M.A.: Uncooled thermal camera calibration and optimization of the photogrammetry process for UAV applications in agriculture. Sensors **17**(10), 2173 (2017)
24. Rolski, T., Schmidli, H., Schmidt, V., Teugels, J.L.: Stochastic Processes for Insurance and Finance, vol. 505. Wiley, New York (2009)

25. Rosenblatt, F.: The perceptron: A probabilistic model for information storage and organization in the brain. Psychol. Rev. **65**(6), 386 (1958)
26. Simsekli, U., Sagun, L., Gurbuzbalaban, M.: A tail-index analysis of stochastic gradient noise in deep neural networks. In: International Conference on Machine Learning, pp. 5827–5837. PMLR (2019)
27. Zhang, H., Wang, D.: Management of postharvest deficit irrigation of peach trees using infrared canopy temperature. Vadose Zone J. **12**(3), vzj2012–0093 (2013)
28. Zhang, H., Wang, D., Ayars, J.E., Phene, C.J.: Biophysical response of young pomegranate trees to surface and sub-surface drip irrigation and deficit irrigation. Irrig. Sci. **35**(5), 425–435 (2017)
29. Zhao, T., Chen, Y., Ray, A., Doll, D.: Quantifying almond water stress using unmanned aerial vehicles (UAVs): Correlation of stem water potential and higher order moments of non-normalized canopy distribution. In: ASME 2017 International Design Engineering Technical Conferences and Computers and Information in Engineering Conference. American Society of Mechanical Engineers (2017)
30. Zhao, T., Koumis, A., Niu, H., Wang, D., Chen, Y.: Onion irrigation treatment inference using a low-cost hyperspectral scanner. In: Multispectral, Hyperspectral, and Ultraspectral Remote Sensing Technology, Techniques and Applications VII. International Society for Optics and Photonics (2018)
31. Zhao, T., Niu, H., Anderson, A., Chen, Y., Viers, J.: A detailed study on accuracy of uncooled thermal cameras by exploring the data collection workflow. In: Autonomous Air and Ground Sensing Systems for Agricultural Optimization and Phenotyping III. International Society for Optics and Photonics (2018)
32. Zhao, T., Yang, Y., Niu, H., Wang, D., Chen, Y.: Comparing U-Net convolutional network with mask R-CNN in the performances of pomegranate tree canopy segmentation. In: Multispectral, Hyperspectral, and Ultraspectral Remote Sensing Technology, Techniques and Applications VII, vol. 10780, p. 107801J. International Society for Optics and Photonics (2018)

Chapter 10
Scale-Aware Pomegranate Yield Prediction Using UAV Imagery and Machine Learning

Abstract Monitoring the development of trees and accurately estimating the yield are important to improve orchard management and production. Growers need to estimate the yield of trees at the early stage to make smart decisions for field management. However, methods to predict the yield at the individual tree level are currently not available due to the complexity and variability of each tree. This study aimed to evaluate the performance of an unmanned aerial vehicle (UAV)-based remote sensing system and machine learning (ML) approaches for a tree-level pomegranate yield estimation. Lightweight sensors, such as the multispectral camera, were mounted on the UAV platform to acquire high-resolution images. Eight characteristics were extracted, including the normalized difference vegetation index (NDVI), the green normalized vegetation index (GNDVI), the RedEdge normalized difference vegetation index (NDVIre), RedEdge triangulated vegetation index (RTVIcore), individual tree canopy size, the modified triangular vegetation index (MTVI2), the chlorophyll index-green (CIg), and the chlorophyll index-rededge (CIre). First, direct correlations were made and the correlation coefficient (R^2) was determined between these vegetation indices and tree yield. Then, machine learning approaches were applied with the extracted features to predict the yield at the individual tree level. The results showed that the decision tree classifier had the best prediction performance, with an accuracy of 85%. The study demonstrated the potential of using UAV-based remote sensing methods, coupled with ML algorithms, for estimating the pomegranate yield. Predicting the yield at the individual tree level will enable stakeholders to manage the orchard on different scales, thus improving field management efficiency.

10.1 Introduction

Due to the recurring water shortages in California, many growers started growing crops that have drought resistance and high economic value to some extent, such as pomegranate [35]. There is approximately 11,000 ha of pomegranate in California, and evidence suggests that pomegranate trees have strong adaptability to a wide range of soil conditions and climates [12, 30]. Research results show that

H. Niu, Y. Chen, *Smart Big Data in Digital Agriculture Applications*, Agriculture Automation and Control, https://doi.org/10.1007/978-3-031-52645-9_10

pomegranate has great potential for human disease treatment and prevention, such as cancer [15, 27]. Pomegranate yield estimation can provide critical information for stakeholders and help them make better decisions on field operations. Therefore, predicting efficient pomegranate yield is economically important in pomegranate production.

The yield of field and woody crops is usually determined by their genotype and environmental conditions, such as soil, irrigation management, weather conditions, etc., making the yield prediction complicated and inaccurate [7, 34]. Thus, many researchers have been working on the yield prediction using a plethora of approaches [6, 26, 31, 36, 37]. For example, [37] developed statistical models using the stochastic gradient boosting method for early and mid-season yield prediction of almond in the central valley of California. Multiple variables were extracted from the remote sensing images, such as canopy cover percentage (CCP) and vegetation indices (VIs). Research results demonstrated the potential of automatic almond yield prediction at the individual orchard level. In [32], Yang et al. estimated the corn yield by using the hyperspectral imagery and convolutional neural networks (CNNs). Results showed that the spectral and color image-based integrated CNN model has a classification accuracy of 75% for corn yield prediction.

Recently, unmanned aerial vehicles (UAVs) and lightweight payloads have been used as a reliable remote sensing platform for many researchers to monitor the crop status temporally and spatially [19, 22, 28, 29]. Equipped with lightweight payloads, such as an RGB camera, a multispectral camera, and a thermal camera, the UAV-based remote sensing system can provide low-cost and high-resolution images for data analysis. For example, in [31], Yang et al. proposed an efficient CNN for rice grain yield estimation. A fixed-wing UAV was adopted to collect RGB and multispectral images to derive the vegetation indices. The results showed that the CNNs trained with RGB and multispectral imagery had better performance than the VIs-based regression model. In [26], Stateras et al. defined the geometry of olive tree configurations and developed a forecasting model of annual production in a non-linear olive grove. Digital terrain model (DTM) and digital surface model (DSM) were generated with high-resolution multispectral imagery. Results showed that the forecasting model could predict the olive yield in kilograms per tree.

However, few studies have investigated the correlation between tree canopy characteristics and yield prediction at the individual tree level. Thus, this chapter aims to estimate the pomegranate tree yield with ten different tree canopy characteristics, which are normalized difference vegetation index (NDVI), green normalized vegetation index (GNDVI), RedEdge normalized difference vegetation index (NDVIre), RedEdge triangulated vegetation index (RTVIcore), canopy size, canopy temperature, irrigation level, the modified triangular vegetation index (MTVI2), the chlorophyll index-green (CIg), and the chlorophyll index-rededge (CIre). For example, NDVI has been commonly used for vegetation monitoring, such as water stress detection [38], crop yield evaluation [39], and evapotranspiration (ET) estimation [20]. The NDVI value is a standardized method to measure healthy vegetation. When the NDVI is high, it indicates that the vegetation has a higher level of photosynthesis. In [7], Feng et al. demonstrated that NDVI and yield had a Pearson correlation coefficient of 0.80. GNDVI and yield had a correlation of 0.53.

The objectives of this chapter were (1) to investigate the correlation between the pomegranate yield and eight different features extracted from high-resolution UAV images and (2) to establish a scale-aware yield prediction model using machine learning approaches. Estimating the yield with scale-aware models will help stakeholders make better decisions for field management at the block or orchard levels.

10.2 Materials and Methods

10.2.1 Experimental Field and Ground Data Collection

This study was conducted in a pomegranate research field at the USDA-ARS, San Joaquin Valley Agricultural Sciences Center (36.594 °N, 119.512 °W), Parlier, California, 93648, USA. The soil types are Hanford fine sandy loam (coarse-loamy, mixed, and thermic Typic Xerorthents). The San Joaquin Valley has a Mediterranean climate with hot and dry summers. Rainfall is insignificant during the growing season, and irrigation is the only source of water for pomegranate growth [30]. Pomegranate (Punica granatum L., cv "Wonderful") was planted in 2010 with a 5 m spacing between rows and a 2.75 m within-row tree spacing in a 1.3 ha field [35].

The pomegranate field was randomly designed into 16 equal blocks, with four replications, to test four irrigation levels (Fig. 10.1). The irrigation volumes were 35%, 50%, 75%, and 100% of crop evapotranspiration (ET_c), measured by the weighing lysimeter in the field. There were five yield sampling trees in each block, 80 sampling trees in total, marked with red labels in Fig. 10.2.

10.2.2 UAV Platform and Imagery Data Acquisition

The UAV-based remote sensing system consisted of a UAV platform, called "Hover," and a multispectral camera (RedEdge M, Micasense, Seattle, WA, USA). The RedEdge M has five different bands, which are blue (B, 475 nm), green (G, 560 nm), red (R, 668 nm), red edge (RedEdge, 717 nm), and near-infrared (NIR, 840 nm). With a Downwelling Light Sensor (DLS), a five-band light sensor that connects to the camera, the RedEdge M can measure the ambient light during a flight mission for each of the five bands. Then, it can record the light information in the metadata of the images captured by the camera. After camera calibration, the information detected by DLS can be used to correct lighting changes during a flight, such as changes in cloud cover during a UAV flight.

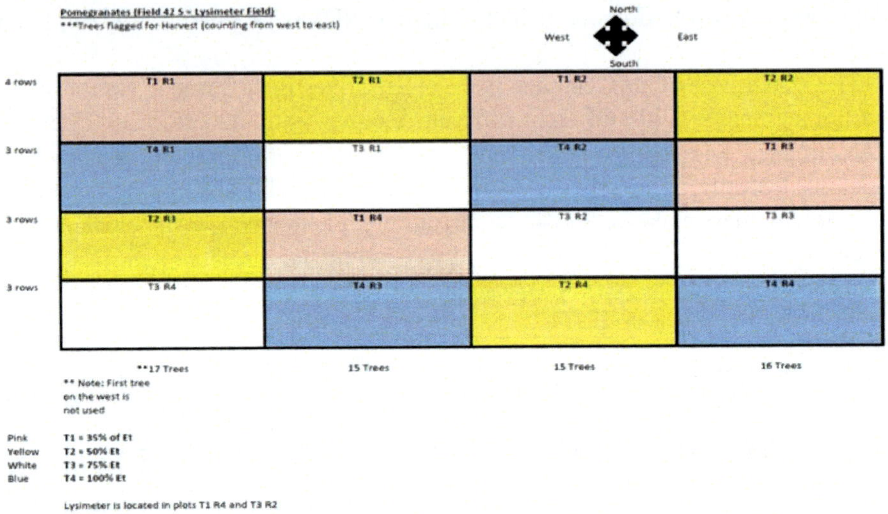

Fig. 10.1 The pomegranate field was randomly designed into 16 equal blocks, with four replications, to test four irrigation levels. The irrigation volumes are 35%, 50%, 75%, and 100% of ET_c, which were measured using the weighing lysimeter in the field

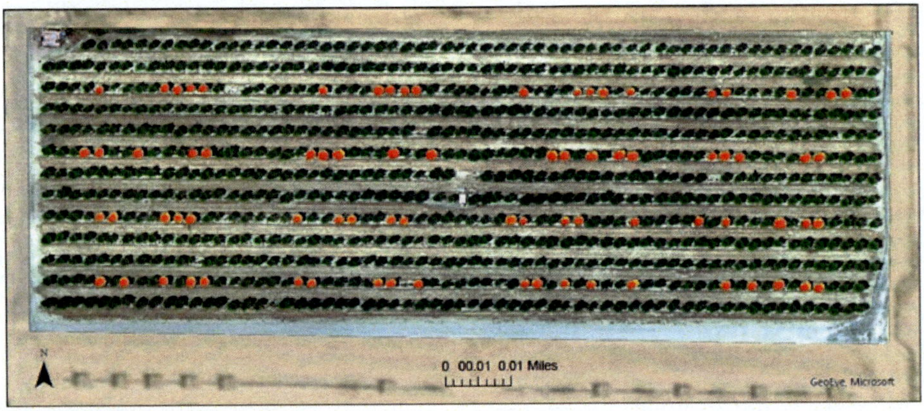

Fig. 10.2 There were five sampling trees in each block, 80 sampling trees in total, marked with red labels

A Mission Planner software was used to design flight missions. The flight height was designed as 60 m above ground level (AGL). UAV image overlap was designed as 75% forward and 70% sideward to stitch UAV images successfully using Agisoft Metashape (Agisoft LLC, Russia).

Table 10.1 UAV image features used in this study

Features	Equations	Related traits	References
NDVI	Eq. (10.1)	Yield, leaf chlorophyll content, biomass	[2, 7, 24]
GNDVI	Eq. (10.2)	Yield, leaf chlorophyll content, biomass	[7, 33]
NDVIre	Eq. (10.3)	Nitrogen, yield	[18, 25]
RTVIcore	Eq. (10.4)	Lear area index, biomass, nitrogen	[13, 17]
MTVI2	Eq. (10.5)	Leaf chlorophyll content	[3, 11]
CIg	Eq. (10.6)	Yield, leaf chlorophyll content	[5, 10]
CIre	Eq. (10.7)	Yield, leaf chlorophyll content	[4, 23]

10.2.3 UAV Image Feature Extraction

The orthomosaic image was used to extract the features of the image defined in Table 10.1. Seven image features were extracted from the multispectral orthomosaic image acquired by the UAV platform. All vegetation indices or features have been commonly used to monitor plant health, nitrogen, biomass, and yield estimation [2, 3, 10, 17, 25, 33].

10.2.3.1 The Normalized Difference Vegetation Index (NDVI)

The NDVI has been commonly used for vegetation monitoring, such as water stress detection [38], crop yield assessment [39], and ET estimation [21]. The value of NDVI is a standardized method to measure healthy vegetation, allowing to generate an image displaying greenness (relative biomass). The NDVI takes advantage of the contrast of the characteristics of two bands, which are the chlorophyll pigment absorptions in the red band (R) and the high reflectivity of plant materials in the near-infrared (NIR) band. When the NDVI is high, it indicates that the vegetation has a higher level of photosynthesis. The NDVI is usually calculated by

$$NDVI = \frac{NIR - R}{NIR + R}, \tag{10.1}$$

where NIR and R are the reflectance of near-infrared and red bands, respectively.

10.2.3.2 The Green Normalized Difference Vegetation Index (GNDVI)

The GNDVI is commonly used to estimate photosynthetic activity and determine water and nitrogen uptake into the plant canopy [7, 33]. The GNDVI is calculated by

$$GNDVI = \frac{NIR - G}{NIR + G}, \tag{10.2}$$

where G stands for the reflectance of the green band.

10.2.3.3 The RedEdge Normalized Difference Vegetation Index (NDVIre)

The NDVIre is a method for estimating vegetation health using the RedEdge band. The chlorophyll concentration is usually higher at the late stages of plant growth; the NDVIre can then be used to map the within-field variability of nitrogen foliage to help better understand the fertilizer requirements of crops [18, 25]. The NDVIre is calculated by

$$NDVIre = \frac{NIR - RedEdge}{NIR + RedEdge}, \tag{10.3}$$

where $RedEdge$ is the reflectance of the RedEdge band.

10.2.3.4 The RedEdge Triangulated Vegetation Index (RTVIcore)

The RTVIcore is usually used for estimating the leaf area index and biomass [13, 17]. It uses the reflectance in the NIR, RedEdge, and G spectral bands, calculated by

$$100(NIR - RedEdge) - 10(NIR - G). \tag{10.4}$$

10.2.3.5 The Modified Triangular Vegetation Index (MTVI2)

The MTVI2 method usually detects the leaf chlorophyll content at the canopy scale, which is relatively insensitive to the leaf area index [3]. MTVI2 uses the reflectance in the G, R, and NIR bands, calculated by

$$\frac{1.5[1.2(NIR - G) - 2.5(R - G)]}{\sqrt{(2NIR + 1)^2 - (6NIR - 5\sqrt{R}) - 0.5}}. \tag{10.5}$$

10.2.3.6 The Green Chlorophyll Index (CIg)

The CIg is for estimating the chlorophyll content in leaves using the ratio of the reflectivity in the NIR and G bands [10], which is calculated by

$$\frac{NIR}{G - 1}. \tag{10.6}$$

10.2.3.7 The RedEdge Chlorophyll Index (CIre)

The CIre is for estimating the chlorophyll content in leaves using the ratio of the reflectivity in the NIR and RedEdge bands [10], which is calculated by

$$\frac{NIR}{RedEdge - 1}. \tag{10.7}$$

10.2.4 The Machine Learning Methods

Several ML classifiers were adopted to evaluate the performance of pomegranate yield estimation, such as "Random Forest" [1], "AdaBoost" [14], "Nearest Neighbors" [9], and "Decision Tree" [16]. The "Random Forest" classifier is a meta-estimator that fits several decision tree classifiers on various sub-samples of the dataset and adopts averaging to improve the predictive accuracy and control overfitting. An "AdaBoost" classifier is also a meta-estimator that begins by fitting a classifier on the original dataset and then fits additional copies of the classifier on the same dataset but where the weights of incorrectly classified instances are adjusted such that subsequent classifiers focus more on complex cases.

The "Nearest Neighbors" method is to apply a predefined number of training samples closest in the distance to the new point and predict the label from these. The samples can be a constant k-nearest neighbor learning or vary based on the local density of points (radius-based neighbor learning). Despite its simplicity, the nearest neighbor's method has been successfully applied for many research problems, such as the handwritten digit classification. As a non-parametric method, it is often successful in classification situations where the decision boundary is very irregular.

The "Decision Trees" are also non-parametric supervised learning methods commonly adopted for classification problems. The objective is to create a model that predicts the value of a target variable by learning simple decision rules inferred from the data features. A tree can be seen as a piecewise constant approximation. "Decision Trees" usually use a white box model, which means the explanation for the condition is easily explained by Boolean logic if a given situation is observable in a model. In contrast, results may be more challenging to interpret for a black box model, such as an artificial neural network.

10.3 Results and Discussion

10.3.1 The Pomegranate Yield Performance in 2019

The pomegranate fruit was harvested from 80 sampling trees in 2019. As mentioned earlier, there were four different irrigation levels in the field, 35%, 50%, 75%, and 100% of ET. The authors then calculated the total fruit weight per tree (kg) and drew the boxplot for each irrigation level (Fig. 10.3). For the 35% irrigation treatment, the total fruit weight per tree was 23.92 kg, which produced the lowest yield. For the 50% irrigation treatment, the total fruit weight per tree was 27.63 kg. For 75% and 100% irrigation treatment, the total fruit weight per tree was 29.84 kg and 34.85 kg, respectively. The pomegranate yield performance at the USDA is consistent with previous research work [35]. Since the authors have the yield data for each sampling tree, machine learning algorithms were used for individual tree-level yield estimation with the eight image features mentioned earlier.

10.3.2 The Correlation Between the Image Features and Pomegranate Yield

Before the vegetation indices were used as input features for ML algorithms, the authors first investigated the correlation (R^2) between the vegetation index and the

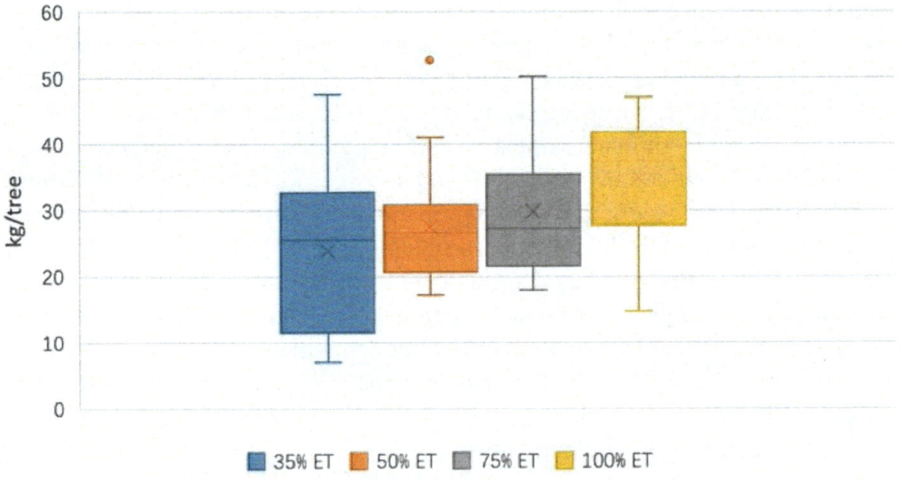

Fig. 10.3 The pomegranate yield performance at the individual tree level in 2019. For the 35% irrigation treatment, the total fruit weight per tree was 23.92 kg, which produced the lowest yield. For the 50% irrigation treatment, the total fruit weight per tree was 27.63 kg. For 75% and 100% irrigation treatment, the total fruit weight per tree was 29.84 kg and 34.85 kg, respectively

pomegranate yield (Fig. 10.4). Each dot represented a mean value of 20 sampling trees. According to the research results, the NDVIre and CIre had relatively higher R^2, which were 0.6963 and 0.6772, respectively. Research results showed that the NDVI and the pomegranate yield had an R^2 of 0.6273. The GNDVI and the yield had an R^2 of 0.5166. The MTVI2 and CIg had an R^2 of 0.4293 and 0.5059, respectively. The RTVIcore had the lowest R^2 of 0.1216. The canopy size had an R^2 of 0.6192. These findings emphasized the importance of yield estimation using vegetation indices. Several researchers reported that vegetation indices could be used for yield estimation [2–5, 7, 17, 24, 25, 33]. The performance of ML algorithms on yield prediction is discussed in the following section.

10.3.3 The ML Algorithm Performance on Yield Estimation

The pomegranate yield data (80 sampling trees) were distributed as 75% for training and 25% for testing using the *train_test_split* method. Considering the dataset was relatively small, the authors used K-fold cross-validation, splitting the training dataset into K folds, then making predictions, and evaluating each fold using an ML model trained on the remaining folds [8]. The classes were defined as low yield and high yield for yield prediction based on a threshold value of 25 kg per tree. For evaluating the trained models, a confusion matrix was used to compare the performances of different classifiers. A confusion matrix was a summary of prediction results on a classification problem. Correct and incorrect predictions were tallied with count values and divided into classes. The confusion matrix provided insight into not only the errors being made by a classifier but also, more importantly, the types of errors that were being made. "True label" meant the ground truth of the yield. "Predicted label" identified the individual tree yield predicted by the trained model.

The trained ML classifiers had distinct test performance for individual tree-level yield prediction. The "Decision Trees" classifier had the highest accuracy of 0.85. Table 10.2 and Fig. 10.5 show the details of the "Decision Trees" method, a non-parametric supervised learning method commonly adopted for classification problems. The objective was to create an ML model that predicted the value of a target variable by learning simple decision rules inferred from the data features (Fig. 10.5). A tree can be seen as a piecewise constant approximation. "Decision Trees" usually uses a white box model, which means the explanation for the condition is easily explained by Boolean logic if a given situation is observable in a model. As shown in Fig. 10.5, the "Decision Trees" ML model started at the root node, and if the NDVIre value were less than 0.334, the prediction process would move to the leaf child node. In this case, the model would predict that the input was a low-yield pomegranate tree. A node's gini attribute measures its impurity: A node is "pure (gini = 0)" if all the training instances it applies are from the same class.

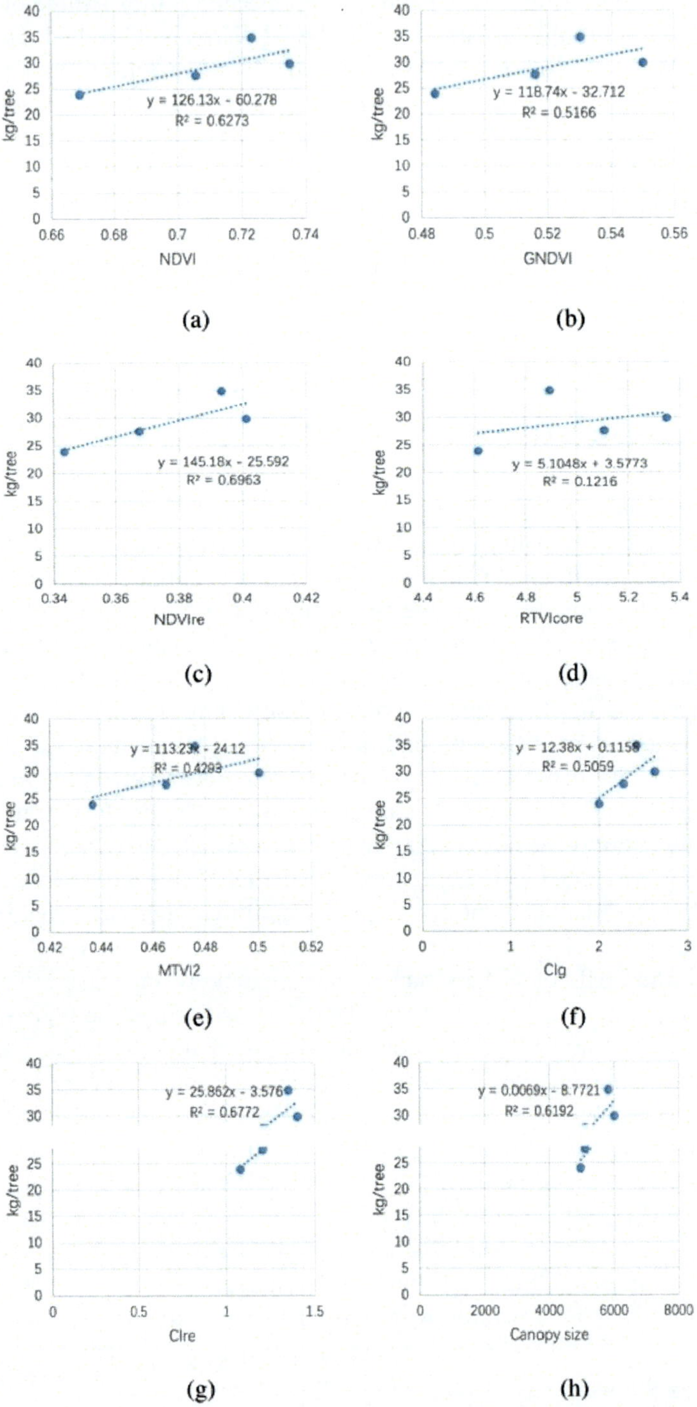

Fig. 10.4 The correlation between the vegetation indices and yield. (**a**) NDVI. (**b**) GNDVI. (**c**) NDVIre. (**d**) RTVIcore. (**e**) MTVI2. (**f**) CIg. (**g**) CIre. (**h**) Canopy size

Table 10.2 The "Decision Tree" performance on yield prediction. "NA" stands for "Not available"

Yield prediction	Precision	Recall	F1-score
Low yield	0.92	0.85	0.88
High yield	0.75	0.86	0.80
Accuracy	NA	NA	**0.85**
Macro avg	0.83	0.85	0.84
Weighted avg	0.86	0.85	0.85

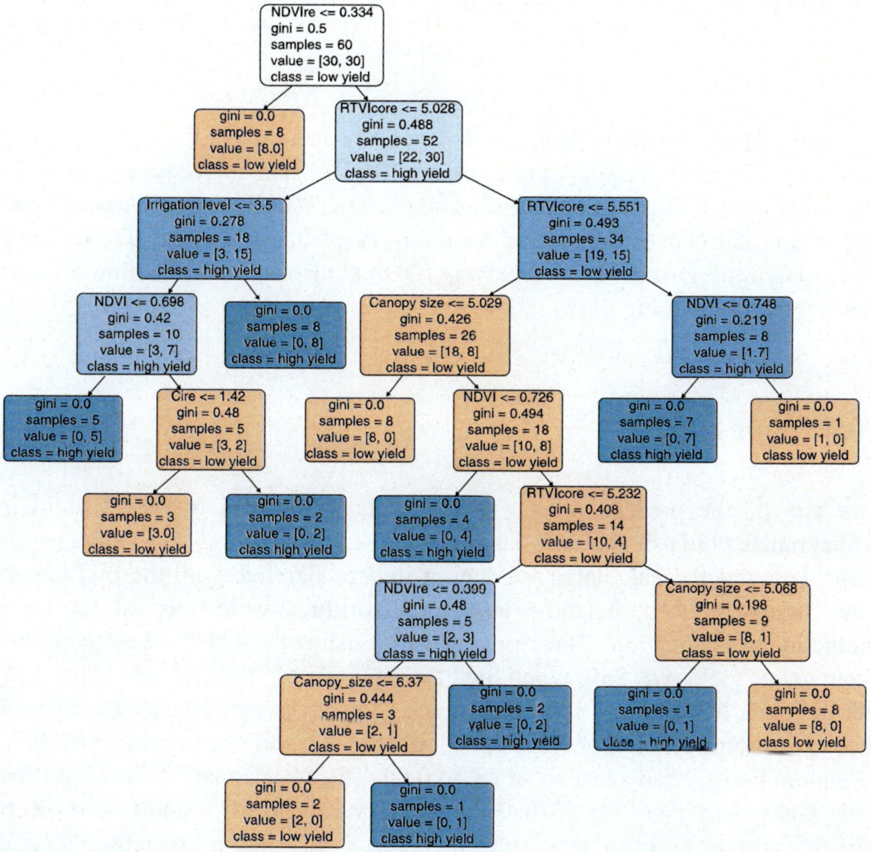

Fig. 10.5 The "Decision Trees" method training process. The "Decision Trees" usually uses a white box model, which means the explanation for the condition is easily explained by Boolean logic if a given situation is observable in a model. As shown here, the "Decision Trees" ML model started at the root node, and if the NDVIre value were less than 0.334, the prediction process would move to the leaf child node. In this case, the model would predict that the input was a low-yield pomegranate tree. A node's gini attribute measures its impurity: A node is "pure (gini = 0)" if all the training instances it applies are from the same class

Table 10.3 The performance of ML methods on yield prediction

Classification methods	Prediction accuracy
"Decision Trees"	**0.85**
"Nearest Neighbors"	0.80
"Support Vector Machine"	0.70
"Random Forest"	0.65
"AdaBoost"	0.80
"Gaussian Process"	0.75
"Gaussian Naive Bayes"	0.60

The bold value is the best model's prediction accuracy.

For the other classifiers' test performance, the accuracy of the k-nearest neighbor was 0.8. "Support Vector Classification (SVC)" had an accuracy of 0.7. The "Random Forest" had a test accuracy of 0.65. The "AdaBoost," "Gaussian Process," and "Gaussian Naive Bayes" had an accuracy of 0.8, 0.75, and 0.6, respectively. The "Quadratic Discriminant Analysis (QDA)" also had a prediction accuracy of 0.8 (Table 10.3 and Fig. 10.6).

10.4 Summary

The aim of the present chapter is to predict individual tree-level yield in the pomegranate field using a UAV-based remote sensing method. The authors collected yield data and the calculated vegetation indices derived from the high-resolution UAV imagery. Then, machine learning algorithms were adopted for the yield prediction classification. The research results showed that the best classification accuracy of yield was 85% when the "Decision Trees" method was being adopted. For the other ML models' test performance, the accuracy of the k-nearest neighbor was 0.8. "Support Vector Classification (SVC)" had an accuracy of 0.7. The "Random Forest" had a test accuracy of 0.65. The "AdaBoost," "Gaussian Process," and "Gaussian Naive Bayes" had an accuracy of 0.8, 0.75, and 0.6, respectively. The "Quadratic Discriminant Analysis (QDA)" also had a prediction accuracy of 0.8. The pomegranate yield information could be reflected by vegetation index data. The research results supported the idea that vegetation indices could be used for yield estimation. Furthermore, the findings of this research provided insights for the scale-aware yield prediction using phenotyping and machine learning technology.

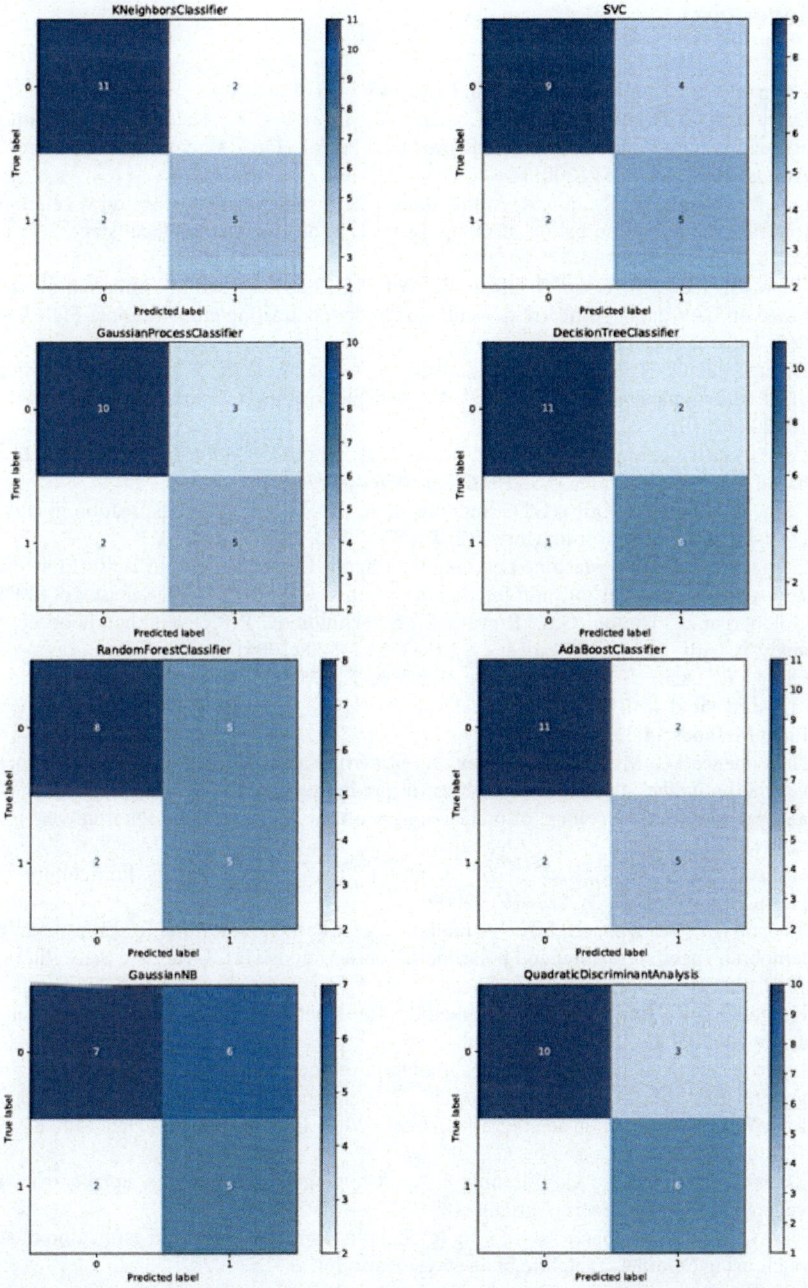

Fig. 10.6 The comparison of the eight different ML classifiers on individual tree-level yield prediction. "True label" meant the ground truth of the yield. "Predicted label" identified the individual tree yield predicted by the trained model. The value 0 meant the low yield; value 1 meant the high yield

References

1. Breiman, L.: Random forests. Mach. Learn. **45**(1), 5–32 (2001)
2. Dalezios, N., Domenikiotis, C., Loukas, A., Tzortzios, S., Kalaitzidis, C.: Cotton yield estimation based on NOAA/AVHRR produced NDVI. Phys. Chem. Earth B Hydrol. Oceans Atmos. **26**(3), 247–251 (2001)
3. Din, M., Zheng, W., Rashid, M., Wang, S., Shi, Z.: Evaluating hyperspectral vegetation indices for leaf area index estimation of Oryza sativa L at diverse phenological stages. Front. Plant Sci. **8**, 820 (2017)
4. Duan, B., Fang, S., Gong, Y., Peng, Y., Wu, X., Zhu, R.: Remote estimation of grain yield based on UAV data in different rice cultivars under contrasting climatic zone. Field Crops Res. **267**, 108,148 (2021)
5. Duan, B., Fang, S., Zhu, R., Wu, X., Wang, S., Gong, Y., Peng, Y.: Remote estimation of rice yield with unmanned aerial vehicle (UAV) data and spectral mixture analysis. Front. Plant Sci. **10**, 204 (2019)
6. Feng, A., Zhang, M., Sudduth, K.A., Vories, E.D., Zhou, J.: Cotton yield estimation from UAV-based plant height. Trans. ASABE **62**(2), 393–404 (2019)
7. Feng, A., Zhou, J., Vories, E.D., Sudduth, K.A., Zhang, M.: Yield estimation in cotton using UAV-based multi-sensor imagery. Biosyst. Eng. **193**, 101–114 (2020)
8. Géron, A.: Hands-on Machine Learning with Scikit-Learn, Keras, and TensorFlow: Concepts, Tools, and Techniques to Build Intelligent Systems. O'Reilly Media, Sebastopol (2019)
9. Goldberger, J., Hinton, G.E., Roweis, S., Salakhutdinov, R.R.: Neighbourhood components analysis. Adv. Neural Inf. Process. Syst. **17**, 513–520 (2004)
10. Gong, Y., Duan, B., Fang, S., Zhu, R., Wu, X., Ma, Y., Peng, Y.: Remote estimation of rapeseed yield with unmanned aerial vehicle (UAV) imaging and spectral mixture analysis. Plant Methods **14**(1), 1–14 (2018)
11. Haboudane, D., Miller, J.R., Pattey, E., Zarco-Tejada, P.J., Strachan, I.B.: Hyperspectral vegetation indices and novel algorithms for predicting green LAI of crop canopies: Modeling and validation in the context of precision agriculture. Remote Sens. Environ. **90**(3), 337–352 (2004)
12. Holland, D., Hatib, K., Bar-Ya'akov, I., et al.: 2 pomegranate: botany, horticulture, breeding. Horticultural Rev. **35**(2), 127–191 (2009)
13. Khosravi, I., Alavipanah, S.K.: A random forest-based framework for crop mapping using temporal, spectral, textural and polarimetric observations. Int. J. Remote Sens. **40**(18), 7221–7251 (2019)
14. Kingma, D.P., Ba, J.: Adam: A method for stochastic optimization. Preprint (2014). arXiv:1412.6980
15. Lansky, E.P., Newman, R.A.: Punica granatum (pomegranate) and its potential for prevention and treatment of inflammation and cancer. J. Ethnopharmacol. **109**(2), 177–206 (2007)
16. Loh, W.Y.: Classification and regression trees. Wiley Interdiscipl. Rev. Data Min. Knowl. Disc. **1**(1), 14–23 (2011)
17. Magney, T.S., Eitel, J.U., Vierling, L.A.: Mapping wheat nitrogen uptake from rapideye vegetation indices. Precis. Agricult. **18**(4), 429–451 (2017)
18. Narin, O.G., Abdikan, S.: Monitoring of phenological stage and yield estimation of sunflower plant using Sentinel-2 satellite images. Geocarto Int., 1–15 (2020)
19. Niu, H., Hollenbeck, D., Zhao, T., Wang, D., Chen, Y.: Evapotranspiration estimation with small UAVs in precision agriculture. Sensors **20**(22), 6427 (2020)
20. Niu, H., Wang, D., Chen, Y.: Estimating actual crop evapotranspiration using deep stochastic configuration networks model and UAV-based crop coefficients in a pomegranate orchard. In: Autonomous Air and Ground Sensing Systems for Agricultural Optimization and Phenotyping V. International Society for Optics and Photonics (2020)
21. Niu, H., Wang, D., Chen, Y.: Estimating crop coefficients using linear and deep stochastic configuration networks models and UAV-based normalized difference vegetation index

(NDVI). In: Proceedings of the 2020 International Conference on Unmanned Aircraft Systems (ICUAS), pp. 1485–1490. IEEE (2020)
22. Niu, H., Zhao, T., Wei, J., Wang, D., Chen, Y.: Reliable tree-level evapotranspiration estimation of pomegranate trees using lysimeter and UAV multispectral imagery. In: 2021 IEEE Conference on Technologies for Sustainability (SusTech), pp. 1–6. IEEE (2021)
23. Peng, Y., Li, Y., Dai, C., Fang, S., Gong, Y., Wu, X., Zhu, R., Liu, K., et al.: Remote prediction of yield based on LAI estimation in oilseed rape under different planting methods and nitrogen fertilizer applications. Agricult. Forest Meteorol. **271**, 116–125 (2019)
24. Ren, J., Chen, Z., Zhou, Q., Tang, H.: Regional yield estimation for winter wheat with MODIS-NDVI data in Shandong, China. Int. J. Appl. Earth Observ. Geoinf. **10**(4), 403–413 (2008)
25. Schwalbert, R.A., Amado, T.J., Nieto, L., Varela, S., Corassa, G.M., Horbe, T.A., Rice, C.W., Peralta, N.R., Ciampitti, I.A.: Forecasting maize yield at field scale based on high-resolution satellite imagery. Biosyst. Eng. **171**, 179–192 (2018)
26. Stateras, D., Kalivas, D.: Assessment of olive tree canopy characteristics and yield forecast model using high resolution UAV imagery. Agriculture **10**(9), 385 (2020)
27. Sumner, M.D., Elliott-Eller, M., Weidner, G., Daubenmier, J.J., Chew, M.H., Marlin, R., Raisin, C.J., Ornish, D.: Effects of pomegranate juice consumption on myocardial perfusion in patients with coronary heart disease. Am. J. Cardiol. **96**(6), 810–814 (2005)
28. Swain, K.C., Thomson, S.J., Jayasuriya, H.P.: Adoption of an unmanned helicopter for low-altitude remote sensing to estimate yield and total biomass of a rice crop. Trans. ASABE **53**(1), 21–27 (2010)
29. Turner, D., Lucieer, A., Malenovskỳ, Z., King, D.H., Robinson, S.A.: Spatial co-registration of ultra-high resolution visible, multispectral and thermal images acquired with a micro-UAV over Antarctic moss beds. Remote Sens. **6**(5), 4003–4024 (2014)
30. Wang, D., Ayars, J., Tirado-Corbala, R., Makus, D., Phene, C., Phene, R.: Water and nitrogen management of young and maturing pomegranate trees. In: III International Symposium on Pomegranate and Minor Mediterranean Fruits 1089, pp. 395–401 (2013)
31. Yang, Q., Shi, L., Han, J., Zha, Y., Zhu, P.: Deep convolutional neural networks for rice grain yield estimation at the ripening stage using UAV-based remotely sensed images. Field Crops Res. **235**, 142–153 (2019)
32. Yang, W., Nigon, T., Hao, Z., Paiao, G.D., Fernández, F.G., Mulla, D., Yang, C.: Estimation of corn yield based on hyperspectral imagery and convolutional neural network. Comput. Electron. Agricult. **184**, 106,092 (2021)
33. Yawata, K., Yamamoto, T., Hashimoto, N., Ishida, R., Yoshikawa, H.: Mixed model estimation of rice yield based on NDVI and GNDVI using a satellite image. In: Remote Sensing for Agriculture, Ecosystems, and Hydrology XXI, vol. 11149, p. 1114918. International Society for Optics and Photonics (2019)
34. Zaman, Q., Schumann, A., Percival, D., Gordon, R.: Estimation of wild blueberry fruit yield using digital color photography. Trans. ASABE **51**(5), 1539–1544 (2008)
35. Zhang, H., Wang, D., Ayars, J.E., Phene, C.J.: Biophysical response of young pomegranate trees to surface and sub-surface drip irrigation and deficit irrigation. Irrigat. Sci. **35**(5), 425–435 (2017)
36. Zhang, X., Toudeshki, A., Ehsani, R., Li, H., Zhang, W., Ma, R.: Yield estimation of citrus fruit using rapid image processing in natural background. Smart Agricult. Technol., 100027 (2021)
37. Zhang, Z., Jin, Y., Chen, B., Brown, P.: California almond yield prediction at the orchard level with a machine learning approach. Front. Plant Sci. **10**, 809 (2019)
38. Zhao, T., Doll, D., Chen, Y.: Better almond water stress monitoring using fractional-order moments of non-normalized difference vegetation index. In: 2017 ASABE Annual International Meeting, p. 1. American Society of Agricultural and Biological Engineers (2017)
39. Zhao, T., Wang, Z., Yang, Q., Chen, Y.: Melon yield prediction using small unmanned aerial vehicles. In: Autonomous Air and Ground Sensing Systems for Agricultural Optimization and Phenotyping II, vol. 10218, p. 1021808. International Society for Optics and Photonics (2017)

Part IV
Towards Smart Big Data in Digital Agriculture

Chapter 11
Intelligent Bugs Mapping and Wiping (iBMW): An Affordable Robot-Driven Robot for Farmers

Abstract This chapter presents an innovative solution, the Intelligent Bugs Mapping and Wiping (iBMW) system, designed as an affordable robot-driven robot for farmers. The chapter commences with an introduction, highlighting the critical need for intelligent and cost-effective pest management solutions in agriculture. Existing solutions in the field are discussed, providing context for the development of iBMW. The chapter then delves into the innovation brought by iBMW, detailing its cognitive capabilities for pest population mapping and wiping. Integration of TurtleBot 3 as the "brain", real-time vision processing, and optimal path planning are outlined as key features. Ethical, cultural, and legal considerations in deploying iBMW are addressed, recognizing the importance of responsible technological implementation. Measuring success is a key focus, with criteria such as the temporal and spatial distribution of pest populations, pesticide usage reduction, and the impact on almond yield highlighted. The chapter concludes with a summary, emphasizing the importance of iBMW in revolutionizing pest management practices for farmers.

11.1 Introduction

Navel orangeworm (NOW) is a common pest of almonds in California. According to [19], California almond harvest is worth of more than 5 billion dollars before processing. The NOW can cause around 30% product loss in almonds from direct consumption [9]. First-instar larvae can bore into the nutmeat, and then the larvae can consume most of the nut, which will produce a lot of webbing and frass. Later, larvae damage can also cause fungal infections as soon as hull splits occur. With 2% total damage from NOW, the loss of almond growers can be $158.75 per acre. For those who had 20% total damage, the loss can be as high as $1742.50 per acre, which is 24% of the total profit $7200 per acre. Therefore, NOW is economically important in California, and early detection of pest population distribution in agricultural system is critical to enabling timely interventions and reducing crop yield losses [14]. Each year in California, eradication of pests is not always available due to the fast expanding distribution of the pest or restrictions regarding the tools available to attempt eradication. As mentioned in NOW, they can fly a quarter-mile or more to

H. Niu, Y. Chen, *Smart Big Data in Digital Agriculture Applications*, Agriculture Automation and Control, https://doi.org/10.1007/978-3-031-52645-9_11

find a new host, making it more difficult to map their distribution. Monitoring is able to give some warnings, but the growers have to respond fast. Traditionally, people use pheromone traps to monitor the flights of NOW, but it needs labors counting the number of moths and tracking the data, which is time-consuming and requires a lot of labor effort.

Typically, growers must take four critical steps in their Integrated Pest Management (IPM) program to reduce NOW damage:

1. Winter sanitation to remove mummy nuts
2. Mapping NOW population and timing of moth flights
3. Pesticide applications or biological control
4. Prompt harvest before a third generation of NOW

In this chapter, the authors will focus on the second and third sections. The goal is to develop an intelligent bugs mapping and wiping (iBMW) robot to perform pest population spatial and temporal distribution and "surgical precision spraying" for pest wipeout. The iBMW is an affordable (less than $1000) robot-driven robot, which has a TurtleBot 3 worked as the robot's brain and an unmanned ground vehicle serviced as the work platform. According to the design, the robot will be able to recognize and classify NOW by using deep learning neural networks. Several iBMWs can also work together in the field in swarming mode day and night, so that it can perform temporal and spatial bugs mapping. Then, based on the mapping results, the iBMWs can determine which areas are at greatest risk and whether wiping treatment is needed in those areas.

Being able to target treatment of infestations will result in lower use of toxic insecticide, time saving, reduced production costs, and greater crop yields. Less pesticide usage can also reduce both air and water contamination, so it can reduce the environmental impact. All of these make farming more economically viable and have the potential to provide us with safety, health, and sustainability in food systems for our growing world population.

11.2 Existing Solutions

Current state-of-the-art methods for identifying pest population distribution require farmers to use egg traps, pheromone traps, or degree-day calculations to monitor the first and second generations of navel orangeworm.

Egg traps usually have a mixture of almond meal and almond oil. They will encourage female moths to lay eggs on the traps. The egg traps are usually placed with a density of 1 trap per 5 acres. Farmers or growers usually check the traps twice a week to obtain eggs and determine egg-laying peaks. Pheromone traps are placed mainly in tree canopies at around 6 to 8 feet to map the flight of adult male moths [11]. Farmers need to count the number of moths in the trap and identify peaks in adult activity in the orchard. Although these methods can help prevent almonds from NOW damage to some extent, scouting in the field and collecting data is still very time-consuming. It also requires a lot of labor to finish the field work. Also, counting the number of moths is an empirical job, even experienced farmers may make mistakes. For example, meal moth (Pyralis farinalis) is also attracted to the traps, which looks very similar to navel orangeworm. Another disadvantage is that the traps are point measurement, which can hardly stand for the temporal and spatial distribution for a whole orchard.

After collecting data from egg or pheromone traps, researchers or farmers will make treatment decisions. The first hull split is the most effective timing to apply for insecticide. Spraying the pesticide at the beginning of hull split if eggs are laid on egg traps. Researchers or farmers can also spray pesticides if pheromone traps indicate that the second flight is starting. Commercial spray rigs are usually used in almond orchards, such as D2-40 engine driven or GB-36 power take-off (PTO), manufactured by Air-O-Fan (Reedley, CA, USA) [21]. The Progressive Ag tower is another popular option for spraying applications. Cone-jet nozzles (Teejet Technologies, Wheaton, IL, USA) have also been widely used. The spray speed for those machines is usually set at 2 to 4 mph. Spray applications of insecticides are mostly empirically based on crop phenology and degree-days using a lower threshold of 55°F and an upper threshold of 94°F. Mating disruption, which uses female-moth pheromones to confuse male moths, is another tool used with insecticides together. The cost estimation for mating disruption is $120 to $160 per acre. For insecticide application, it is about $40 to $60 per spraying.

Researchers or farmers typically spray a whole orchard to deal with potential pest risks, for which the insecticide is not used efficiently. The results of the study [3] proved that most of the pesticide was not used on the target trees. Seiber found that 88% of the pesticide was on the orchard floor instead of the trees [20]. Cross also investigated the effects of spray droplet size and application volumes of 10 to 80 gallons per acre on spray deposition in an orchard [2]. The results showed that 43% to 61% of the applied spray was lost to the ground deposit within 15 feet of the row being sprayed. Off-target pesticides result in significant waste of money, as well as environmental issues, such as air pollution and water contamination.

To reduce the use of pesticides and improve efficiency, researchers tried several different new technologies. For example, sensor-equipped sprayers, also called "Smart sprayer," use ultrasonic or optical sensors to detect the presence or absence of trees [7]. Giles found that this technology could save 28% to 34% of pesticides when using this system. The basic idea is to turn on the spray nozzles when the target trees appear in the spray zones [3]. The sprayer will turn off when it is between trees or when the trees are absent. The experiment was designed in a commercial almond orchard near Modesto, CA. Most of the trees were about 20 years old and are planted in a 23-foot diamond pattern. However, in principle, these sprayers are the same as conventional sprayers. They did not consider the temporal and spatial distribution of the pest population. In this chapter, the authors believe iBMWs can improve pesticide spraying efficiency a lot based on the innovations mentioned in the next section.

11.3 iBMW Innovation

As mentioned previously, the iBMW is an intelligent robot which can recognize and classify the NOW by using deep learning neural networks. As shown in Fig. 11.1, by using the swarming mechanism, several iBMWs can scout the orchard together to map the pest population on a large scale. Based on mapping analysis, iBMWs can determine which areas are at greatest risk and whether wiping treatment is necessary. Once the whole system is working, the final goal is to use the iBMW system to monitor more kinds of tree, such as walnuts, pistachios, and pomegranates. Such flexibility can make iBMWs have an impact on more than just one type of field. Based on farmer requirements, the authors will add more agricultural applications, such as soil moisture monitoring, water stress detection [23], and 3D canopies modeling, on this taskable robot platform.

Currently, the objective is to make the iBMWs work perfectly in an almond orchard in Merced (37.493°N, −120.634°W). There are three kinds of almonds: Nonpareil, Carmel, and Monterey, all planted in Lovell peach rootstock, spaced at 5.5 m × 6.1 m, with Rocklin loam and Greenfield sandy soil [23]. In order to achieve pest mapping and wiping, our team has been working in the almond orchard

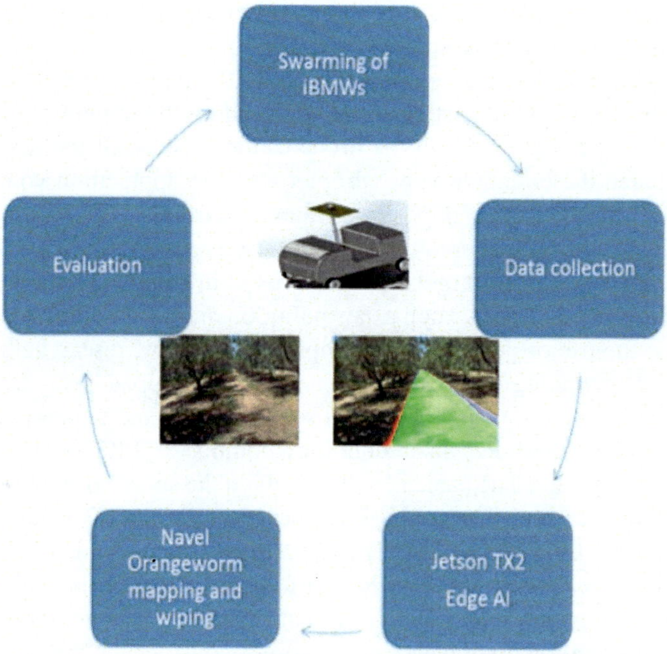

Fig. 11.1 The iBMW workflow. The iBMW is an intelligent robot which can recognize and classify the NOW by using deep learning neural networks. By using swarming mechanism, several iBMWs can scout in orchard together to realize pest population mapping in a large scale. Based on the mapping analysis, the iBMWs can determine which areas are at the greatest risk and whether wiping treatment is needed

several times to collect videos and images to feed into the deep learning algorithms to realize the cognition of the iBMW. We will collect more data in different almond orchards in the future to make iBMWs more robust and adaptive.

11.3.1 Cognitive of Pest Population Mapping and Wiping

Instead of setting up in the almonds' canopies, the egg traps or pheromone traps will be equipped on iBMW's deluxe rod actuator, as shown in Fig. 11.2, which can change the height to six or eight feet when running in the field. Then, iBMWs can scout in orchard together to record the number of moths in the traps by using high-resolution RGB cameras mounted on the iBMWs. High-resolution images enable that eggs and adult moths can be seen clearly in images. Then, deep learning neural networks, such as Faster R-CNN [8] and Fully Convolutional Neural Networks [13], can be used to train image datasets to obtain a NOW detection model. The model will be used to recognize and classify the NOW in the images.

Several iBMWs can work in the field together in swarming mode day and night if necessary. Thus, it can map the NOW in almond orchard temporally and spatially. The high-temporal and spatial resolution images will help farmers determine where,

Fig. 11.2 Concept iBMW

when, and what kind of treatments are needed. For example, the iBMW can pinpoint infestations and problems areas, so that it is unnecessary to treat all areas in a large scale of field. Sprayer system can be mounted on the iBMWs to realize pesticide applications. This will definitely save many hours for field scouting and less pesticide will be used. It will directly or indirectly contribute to agricultural, environmental, and economic sustainability due to minimized application pesticides and improve yield estimation.

The challenges for NOW detection highly depend on how accurate our deep learning algorithms will be. During the image processing, NOW may not be treated as bugs or other pests on traps may be recognized as NOW; both could happen during the training process. We will carefully select training and testing data samples during image processing. Several image processing techniques will be applied to ensure we can get the best accuracy results.

11.3.2 iBMW with TurtleBot 3 as "Brain"

For the iBMW platform, we used a TurtleBot 3 which is operated by ROS (Robot Operating System), as the robot's brain and control the iBMW's behaviors in the field. ROS is a flexible framework for writing robot software, so that it can save us a lot of time on software development. TurtleBot 3 is a low-cost robot platform with open-source software. There are mobile base, 3D sensor, and laptop computer in the TurtleBot 3 kit. TurtleBot 3 has many similar abilities of the large company's robot platforms, such as PR2. With the TurtleBot 3 components, we can even create real-time obstacles avoidance and autonomous navigation.

11.3.3 Real-Time Vision Processing

We will also use the Jetson TX2 to realize cognition mechanism by deep learning algorithms. Field road detection experiment results are shown in Fig. 11.3. Jetson TX2 is the fastest, most power-efficient embedded AI computing device. It is built around an NVIDIA GPU, loaded with 8 GB of memory and 59.7 GB/s of memory bandwidth, which will benefit the iBMWs to react faster in a real-time pest monitoring. Deep learning can allow the iBMW to adapt to field conditions based on computer vision instead of only relying on GPS (GPS sometimes is weak under tree canopies). The iBMW currently can run 3.7 miles per hour. Emergency shut-off can also avoid injury to workers or crops if an object shows up in iBMW's path accidentally. All of these advantages will make the iBMW more time efficient, consistent, and reliable.

iBMW, which includes an artificial intelligence platform, can also enable researchers to conduct proximate sensing and mapping missions without human intervention. Real-time vision processing enabled by cognition algorithms is one

Fig. 11.3 Road detection by Jetson TX2

of the most important factors. Typically, there are three steps of the iBMW movement behaviors, stimulus perception, judgment and decision-making, and stimulus performance. In a recent cognitive algorithm used by robot, it can even recognize specific objects in the environment, by using cognitive algorithms to develop a deeper understanding of the scene based on objects and their spatial relationships.

Reconstruction and mapping Monocular Simultaneous Localization and Mapping (SLAM) can also be applied on iBMW platform. SLAM relies on cognition techniques to calculate the position of the robot and to map the environment [10]. Real-time monocular SLAM was thought not possible because of the high computational cost. However, there are many kinds of SLAM packages now, such as LSD-SLAM [4], RatSLAM [15], and SVO [5]. Researchers compared many different open-source vision-based SLAM packages in a paper at 2016 [12]. It turned out that ORB-SLAM [16] is the best performing package. ORB-SLAM uses ORB features, which are binary features that are invariant to rotation and scale [10]. ORB-SLAM also allows the robots to initialize with no user input, which is important for iBMW platforms as an autonomous robot system. Additionally, when the ORB-SLAM is matching features, the number of key frames increases because of the complexity of the images. This is done by "culling" key frames that are not used and allows for ORB-SLAM to run for longer periods of time [10].

Using deep learning, recognition can bridge the gap between perception and intelligence. For example, perception missions can collect lots of data, but much of it was discarded to simplify interpretation by higher level tasks [22], e.g., a 3D object can become a point in space. In Arne's paper [22], it used a deep learning framework to replace these interfaces with learned interfaces. This allows for error backpropagation that can adapt each module to the robot's task. Deep learning is also a state-of-the-art technique for image processing. It can target land cover and crop type classification from the images collected from sensors and cameras [26]. Safety is another concern by farmers and growers, and the iBMW should have the ability to detect workers and farmers and then will make the right decision in the field.

11.3.4 Optimal Path Planning Enabled by iBMW

The iBMW can also contribute to optimal path planning research. Traditionally, there is not much path planning knowledge for field operations [18]. Researchers execute field operations based on experience. According to a study of spraying vehicles with a positioning system, it showed that 16% of the driving distance could be saved, which expressed a clear need for robot path planning. In the past few years, many algorithms have been developed to realize real-time path planning systems for robots. For example, in [1] , the authors used a visibility graph algorithm and Dijikstra's algorithm to choose optimal path for a robot to traverse in the environment.

Usually, a path planning algorithm needs to meet the following rules [17]:

1. The path planning should have the lowest cost.
2. The resulting path should be fast and accurate.
3. The algorithm should be adaptive with different maps.

The iBMW, as a low-cost robot platform, can be used for basic research of path planning. As mentioned in the previous section, iBMWs use TurtleBot 3 components to control the robot's behavior. We can create real-time obstacles avoidance and autonomous navigation based on the ROS. With the Jetson TX2, the iBMW currently can do emergency shut-off, which avoids injury to workers or crops if an object shows up in iBMW's path accidentally. All of these advantages will make the iBMW more time efficient, consistent, and reliable for path planning research.

11.3.5 Ethical, Cultural, and Legal Matters

With any new practice growers are concerned about if it works and will it save them time and money. Growers would probably worry that there is no time or money saved, as any time or money spent checking traps or setting up pheromone traps would just be spent maintaining machines. This would probably be a big concern, as these are unfamiliar and perceptually more complex machines for growers. We are competing with a large market that has been promising a lot with automation and mechanization over the years and have not really delivered. Thus, growers are getting more skeptical. Getting Bowles Farming Company involved is a good step as we need to get growers to buy-in to get other growers to buy into the concept. We would prefer to spend more time convincing one grower than a full room of growers. If we can convince one person and he or she uses it and it works, they will go about convincing other growers, as well as this naturally happening when a neighbor uses something and it works.

11.4 Measuring Success

There are three outcome indicators that the authors will be monitoring throughout the iBMWs project, the pest population temporal and spatial distribution, the amount of pesticide being used, and the target trees' almond yield.

11.4.1 NOW Population Temporal and Spatial Distribution

The egg traps are usually placed in the orchard in March at a rate of 1 trap per 10 acres. Researchers also hang one pheromone trap per 50 acres and at least two traps per orchard. Farmers usually monitor the traps once a week. The pest population distribution data acquired by using the above method will be used as groudtruth. The authors will compare the iBMW's pest population temporal and spatial distribution with the groudtruth to see if we can get similar or even better results. This will definitely reduce the labor cost, and early detection of the pest will also reduce crop losses made by NOW.

11.4.2 The Amount of Pesticide Being Used

The pesticide usage is another indicator to measure during the growing season. The authors will set up several homogeneous almond blocks. The authors will use commercial spray rigs to spray half of the almond blocks. The iBMWs will be used to spray the rest according to iBMW pest population distribution results. Each spray amount will be recorded during the growing season. The total amount of pesticide for both methods will be calculated after the last spray at the end of growing season. As mentioned in the existing solution section, Giles found that his technology could save 28%–34% of pesticide when using his system [7]. In principle, however, these sprayers are the same with conventional sprayers. They did not consider the pests population temporal and spatial distribution. The authors believe iBMWs can improve pesticide spraying efficiency by reduced use of large-scale pesticide. This low-environmental impact solution will reduce air and water contamination resulting from orchard pesticide usage. Therefore, it can provide us a safe, healthy, and sustainable food systems. The reduction of pesticide cost can be substantial. Based on UC Cooperative publications, the estimate pesticide costs per acre in California orchard crops can be around $250 [6]. Based on [3], their method saved 10%–20% pesticide in Sacramento Valley almonds. The authors will evaluate iBMWs in the following season to see our results.

11.4.3 The Target Trees' Almond Yield

The almond yield will be measured by the researchers or farmers in orchard. The authors will compare the results between conventional method and iBMWs to see if iBMWs have any benefit to the almond yield when applied iBMWs spray system.

11.5 Summary

In general, this chapter proposed an intelligent bugs mapping and wiping (iBMW) robot concept to perform pest population spatial and temporal distribution and "surgical precision spraying" for pest wipeout. Based on the design, the robot will be able to recognize and classify the NOW by using an innovative method for NOW population temporal and spatial distribution. Several iBMWs can also work in the field together in swarming mode. Then, based on the mapping results, the iBMWs can determine which areas are at the greatest risk and whether wiping treatment is needed in those areas.

Being able to target treatment of infestations will result in lower use of toxic insecticide, time saving, reduced production costs, and greater crop yields. Less pesticide usage can also reduce both air and water contamination so that it can reduce the environmental impact. All of these make farming more economically viable and have the potential to provide us safety, health, and sustainability in food systems for our growing world population. Based on the research purpose, the authors can also use this iBMW robot platform for more agricultural applications, such as onion irrigation treatment detection [24], tree canopies segmentation [25, 26], and so on.

References

1. Alexopoulos, C., Griffin, P.M.: Path planning for a mobile robot. IEEE Trans. Syst. Man Cybern. **22**(2), 318–322 (1992)
2. Cross, J., Walklate, P., Murray, R., Richardson, G.: Spray deposits and losses in different sized apple trees from an axial fan orchard sprayer: 1. effects of spray liquid flow rate. Crop Protect. **20**(1), 13–30 (2001)
3. Downey, D., Giles, D., Klassen, P., Niederholzer, F., et al.: "Smart" sprayer technology provides environmental and economic benefits in California orchards. Calif. Agricult. **65**(2), 85–89 (2011)
4. Engel, J., Schöps, T., Cremers, D.: LSD-SLAM: Large-scale direct monocular SLAM. In: European Conference on Computer Vision, pp. 834–849. Springer (2014)
5. Forster, C., Pizzoli, M., Scaramuzza, D.: SVO: Fast semi-direct monocular visual odometry. In: 2014 IEEE International Conference on Robotics and Automation (ICRA), pp. 15–22. IEEE (2014)
6. Freeman, M., Viveros, M., Klonsky, K., De Moura, R.: Sample Costs to Establish an Almond Orchard and Produce Almonds: San Joaquin Valley South Micro-Sprinkler Irrigation. University of California Cooperative Extension, Davis, CA (2008)
7. Giles, D.K., Delwiche, M., Dodd, R.: Control of orchard spraying based on electronic sensing of target characteristics. Trans. ASAE **30**(6), 1624–1636 (1987)
8. Girshick, R.: Fast R-CNN. In: Proceedings of the IEEE International Conference on Computer Vision, pp. 1440–1448 (2015)
9. Higbee, B.S., Siegel, J.P.: Field efficacy and application timing of methoxyfenozide, a reduced-risk treatment for control of navel orangeworm (lepidoptera: Pyralidae) in almond. J. Econ. Entomol. **105**(5), 1702–1711 (2012)
10. Hood, S., Benson, K., Hamod, P., Madison, D., O'Kane, J.M., Rekleitis, I.: Bird's eye view: Cooperative exploration by UGV and UAV. In: 2017 International Conference on Unmanned Aircraft Systems (ICUAS), pp. 247–255. IEEE (2017)
11. Kuenen, L., Siegel, J., et al.: Sticky traps saturate with navel orangeworm in a nonlinear fashion. Calif. Agricult. **70**(1), 32–38 (2016)
12. Li, A.Q., Coskun, A., Doherty, S.M., Ghasemlou, S., Jagtap, A.S., Modasshir, M., Rahman, S., Singh, A., Xanthidis, M., O'Kane, J.M., et al.: Experimental comparison of open source vision-based state estimation algorithms. In: International Symposium on Experimental Robotics, pp. 775–786. Springer (2016)
13. Long, J., Shelhamer, E., Darrell, T.: Fully convolutional networks for semantic segmentation. In: Proceedings of the IEEE Conference on Computer Vision and Pattern Recognition, pp. 3431–3440 (2015)
14. Makori, D., Fombong, A., Abdel-Rahman, E., Nkoba, K., Ongus, J., Irungu, J., Mosomtai, G., Makau, S., Mutanga, O., Odindi, J., et al.: Predicting spatial distribution of key honeybee pests in Kenya using remotely sensed and bioclimatic variables: Key honeybee pests distribution models. ISPRS Int. J. Geo Inf. **6**(3), 66 (2017)
15. Milford, M.J., Wyeth, G.F., Prasser, D.: RatSLAM: A hippocampal model for simultaneous localization and mapping. In: IEEE International Conference on Robotics and Automation, 2004. Proceedings. ICRA'04. 2004, vol. 1, pp. 403–408. IEEE (2004)

16. Mur-Artal, R., Montiel, J.M.M., Tardos, J.D.: ORB-SLAM: A versatile and accurate monocular SLAM system. IEEE Trans. Robot. **31**(5), 1147–1163 (2015)
17. Niederberger, C., Radovic, D., Gross, M.: Generic path planning for real-time applications. In: Proceedings Computer Graphics International, 2004, pp. 299–306. IEEE (2004)
18. Oksanen, T., Visala, A.: Coverage path planning algorithms for agricultural field machines. J. Field Robot. **26**(8), 651–668 (2009)
19. Rovnyak, A.M., Burks, C.S., Gassmann, A.J., Sappington, T.W.: Interrelation of mating, flight, and fecundity in navel orangeworm females. Entomologia ExperimentalisET Applicata **166**(4), 304–315 (2018)
20. Seiber, J.N., Wilson, B.W., McChesney, M.M.: Air and fog deposition residues of four organophosphate insecticides used on dormant orchards in the San Joaquin Valley, California. Environ. Sci. Technol. **27**(10), 2236–2243 (1993)
21. Siegel, J.P., Strmiska, M.M., Niederholzer, F.J., Giles, D.K., Walse, S.S.: Evaluating insecticide coverage in almond and pistachio for control of navel orangeworm (Amyelois transitella)(Lepidoptera: Pyralidae). Pest Manag. Sci. **75**(5), 1435–1442 (2018)
22. Suppé, A.J., Hebert, M.: Using deep learning to bridge the gap between perception and intelligence. In: Unmanned Systems Technology XIX, vol. 10195, p. 1019503. International Society for Optics and Photonics (2017)
23. Zhao, T., Doll, D., Chen, Y.: Better almond water stress monitoring using fractional-order moments of non-normalized difference vegetation index. In: 2017 ASABE Annual International Meeting, p. 1. American Society of Agricultural and Biological Engineers (2017)
24. Zhao, T., Koumis, A., Niu, H., Wang, D., Chen, Y.: Onion irrigation treatment inference using a low-cost hyperspectral scanner. In: Multispectral, Hyperspectral, and Ultraspectral Remote Sensing Technology, Techniques and Applications VII. International Society for Optics and Photonics (2018)
25. Zhao, T., Niu, H., de la Rosa, E., Doll, D., Wang, D., Chen, Y.: Tree canopy differentiation using instance-aware semantic segmentation. In: 2018 ASABE Annual International Meeting. American Society of Agricultural and Biological Engineers (2018)
26. Zhao, T., Yang, Y., Niu, H., Wang, D., Chen, Y.: Comparing U-Net convolutional network with mask R-CNN in the performances of pomegranate tree canopy segmentation. In: Multispectral, Hyperspectral, and Ultraspectral Remote Sensing Technology, Techniques and Applications VII, vol. 10780, p. 107801J. International Society for Optics and Photonics (2018)

Chapter 12
A Non-invasive Stem Water Potential Monitoring Method Using Proximate Sensor and Machine Learning Classification Algorithms

Abstract This chapter introduces a novel method for non-invasive monitoring of stem water potential using proximate sensors and machine learning classification algorithms. The chapter begins with an overview, emphasizing the significance of non-invasive techniques in understanding plant water status for effective agricultural management. The Materials and Methods section provides a detailed description of the walnut study area, the utilization of a radio frequency sensor for reflectance measurements, and the data collection and processing procedures. The implementation of scikit-learn classification algorithms is elucidated as a key component of the monitoring method. The Results and Discussion section presents findings and insights derived from the application of the proposed method. The chapter concludes with a summary, highlighting the potential of this non-invasive approach for stem water potential monitoring. This research opens avenues for advancing digital agriculture through enhanced water management practices. References are provided for further exploration of the methodologies and concepts discussed in this chapter.

12.1 Introduction

Improving the capacity to determine plant water status is critical to solve the drought responses and water requirements of crops and trees [1]. Stem water potential (SWP) is a direct measure of water tension within the plant, implying the water status directly in the plant–soil–atmospheric continuum [3]. SWP also summarizes the effects of both soil moisture and evapotranspiration. These advantages make it a popular index in crops and trees. The details of SWP measurement methods can be referred to [5]. Water mass per leaf area (WMA) and relative water content (RWC) are also critical indices. WMA represents the absolute tissue water content normalized by the leaf area. RWC is normalized by the saturated water content, which provides information about cell volume shrinkage [7]. However, those traditional methods require the excision of tissues and laboratory analysis, either gravimetric in the case of WMA and RWC or using the pressure chamber for SWP [4]. For example, it is labor-intensive and time-consuming to conduct

H. Niu, Y. Chen, *Smart Big Data in Digital Agriculture Applications*, Agriculture Automation and Control, https://doi.org/10.1007/978-3-031-52645-9_12

SWP measurement, requiring at least 10 minutes after the leaf being covered with a foil bag until the water potential is well balanced all around the leaf. Thus, the measurement of plant water status has commonly been limited to small study scales [2].

Compared with the traditional methods, non-invasive methods have great potential for improving ground-based and remote sensing methods in water relations research and their applications in agriculture. In [2], Browne et al. demonstrated a refined method and physical model to predict WMA, RWC, and leaf water potential using terahertz transmission. With the development of unmanned aerial vehicles (UAVs) technology, UAVs have been commonly used in water stress-related research, such as irrigation management [8], evapotranspiration estimation [6], and tree canopy detection [9]. However, UAVs are limited by flight time and payload capability. Such limitations do not apply to proximate sensors. It was the objective of this study to collect data with a pocket-sized, cutting-edge technology radio frequency tridimensional sensor and then implement several scikit-learn classifiers to detect SWP in a walnut orchard.

In this study, the authors applied a low-cost proximate radio frequency tridimensional sensor "Walabot" and machine learning classification algorithms to predict the walnut SWP. Walnut leaves from trees of a different known SWP were placed on this sensor to test if the Walabot can detect small changes in the water stress levels. Hypothetically, waveforms generated by different signals may be useful to classify SWP levels. Scikit-learn classification algorithms, such as "Neural Networks," "Random forest," "Adam optimizer," "Decision Tree," "Support Vector Machine," and "Nearest Neighbors," were applied for data processing and evaluation.

12.2 Materials and Methods

12.2.1 Walnut Study Area

This study was conducted in a walnut orchard at Merced, CA, USA (37.476°N, 120.456°W). There were three types of spacing treatment, 22'×22', 26'×26', and 30'×30'. The region of interest was marked with the red box (Fig. 12.1). To determine the SWP of a walnut tree, midday SWP was measured by using a pressure chamber weekly from May to September 2018.

12.2.2 Reflectance Measurements with a Radio Frequency Sensor

The SWP data were collected with a radio frequency 3D sensor called Walabot (Vayyar Imaging Ltd, Yehud-Monosson, Israel). The Walabot was used to measure

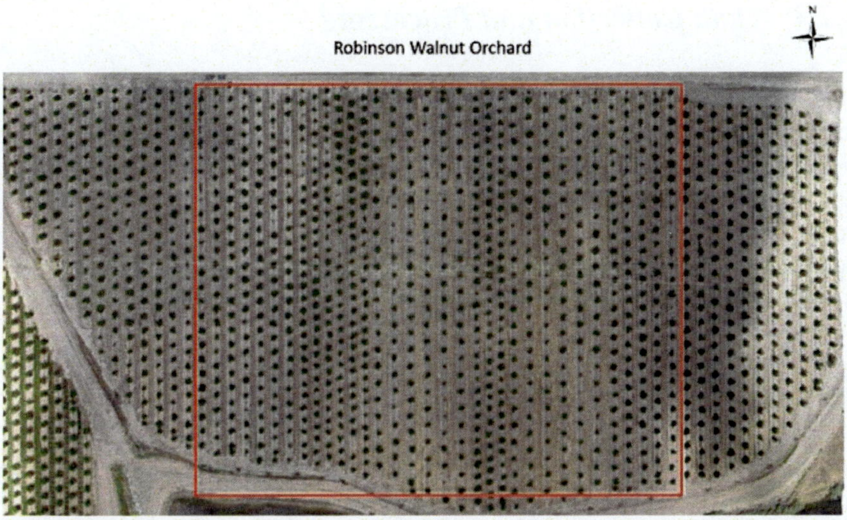

Fig. 12.1 This study was conducted in a walnut orchard at Merced, CA, USA (37.47°N, 120.45°W). There were three types of spacing treatment, 22' × 22', 26' × 26', and 30' × 30'. The region of interest was marked with the red box. To determine the SWP of a walnut tree, midday SWP was measured by using a pressure chamber weekly from May to September 2018

the reflectance of radio frequencies of the walnut leaves. Then, the reflectance was visualized using PyTorch. For each measurement, a time-domain reflectance signal was recorded. Records of the different reflectance strengths were visualized in dataset. The frequency range was 3.3–10 GHz for the US/FCC model and 6.3–8 GHz for the EU/CE model. The average transmitted energy of both models was below 41 dBm/MHz and did not constitute health concerns for the user, nor did it impact the tested leaf tissues.

The Walabot used an antenna array to send radio frequency to the area in front of it and then captured the returning signals (reflectance). The excitation signals were produced, and the reflectance was recorded by integrated circuits. Based on the technical specifications, the Walabot could sense the environment by transmitting, receiving, and recording signals from multiple antennas. Recordings of multiple transmit-receive antenna pairs were analyzed to build a 3D image of the examined environment. Changes in the environment were characterized by analyzing sequences of images. Then, researchers could detect changes in the environment by analyzing the sequences of images. Current commercial applications include the uses of short-range imaging into dielectric environments, for example, drywall and concrete, to detect otherwise invisible objects within these structures.

12.2.3 Data Collection and Processing

In this study, the Walabot was used to measure radio frequency reflectance from the walnut leaves on 8-28-18 and 9-10-18. There were 16 sampling trees, which had unique ID numbers (Table 12.1). For example, "20-4" meant the fourth tree on row 20. Each sampling tree was measured five times to reduce the likelihood of errors or anomalous results. Based on the SWP range, the walnut trees were classified into three levels (Table 12.2).

12.2.4 Scikit-Learn Classification Algorithms

Six different methods in generating classifiers were used to evaluate the detection performance for SWP levels. "Neural Net," "Random Forest," "AdaBoost," "Support Vector Machine," "Nearest Neighbors," and "Decision Trees" were used as classification algorithms. For example, the "Decision Trees" are a non-parametric supervised learning method used for classification and regression. The goal was to create a model that predicted the value of a target variable by learning simple decision rules inferred from the data features.

Table 12.1 The stem water potential of walnut leaves. There were 16 sampling trees, which had unique ID numbers. For example, "20-4" meant the fourth tree on row 20. Each sampling tree was measured five times to reduce the likelihood of errors or anomalous results

Sampling trees	SWP (8-28-2018)	SWP (9-10-2018)
20-4	4.14	5.67
20-12	4.53	4.95
20-20	4.63	4.26
26-4	5.10	4.95
26-12	3.76	4.17
26-20	4.25	4.02
26-28	5.25	4.44
32-4	3.53	4.85
32-12	2.84	1.36
32-20	3.78	5.27
38-4	5.06	5.37
38-12	3.46	3.56
38-20	3.64	4.28
44-4	4.60	4.58
44-12	3.45	4.49
44-20	4.07	4.29

Table 12.2 The SWP levels of walnut leaves. Based on the SWP range, the walnut trees were classified into three levels

SWP levels	Pressure applied (Bar)
Level 1	< 3.5
Level 2	3–5
Level 3	> 5

12.3 Results and Discussion

The collected SWP data were converted into 2048-dimensional vectors for scikit-learn algorithms data processing. The data were distributed as 70% for training and 30% for testing. For evaluating the trained models, a confusion matrix was used to compare performances of different classifiers. A confusion matrix is a summary of prediction results on a classification problem. The number of correct and incorrect predictions is tallied with count values and divided into classes. The scikit-learn classifiers had different levels of performance (Fig. 12.2). The confusion matrix provided insight into not only the errors being made by a classifier but also, more importantly, the types of errors that were being made. "True label" meant the ground truth of SWP levels. "Predicted label" identified the SWP levels predicted by the trained model. The "Decision Tree" had the highest accuracy of 0.78 (Table 12.3). The accuracy of the "Support Vector Machine" was 0.62. The "Random Forest" also had a low prediction accuracy of 0.60. The "AdaBoost," "Nearest Neighbors," and "Neural Network" had an accuracy of 0.60, 0.65, and 0.62, respectively.

In the analysis as "Decision Tree" for the SWP Level 1, the trained model predicted the test data was in the range of Level 3 (Table 12.4). Therefore, the prediction accuracy (F1-score) of Level 1 was zero. The model successfully classified 22 out of 25 samplings in Level 2. For Level 2, the model had performance with an F1-score of 0.92. For Level 3, the model classified nine out of ten samplings. The overall prediction accuracy of the trained model was 0.78. The model had poor performance when the SWP was low. Perhaps the difference between features was minimal when the SWP was at this low level and could not be detected. Another reason might be the performance of the "Decision Tree" classifier. Further study needed to be conducted to find better algorithms for SWP detection. The dataset was also very small. More data will be collected and added to train a more robust model for SWP detection.

Table 12.3 The performance of classification methods. The accuracy of the "Support Vector Machine" was 0.62. The "Random Forest" also had a low prediction accuracy of 0.60. The "AdaBoost," "Nearest Neighbors," and "Neural Network" had an accuracy of 0.60, 0.65, and 0.62, respectively

Classification methods	Prediction accuracy
"Neural Network"	0.62
"Support Vector Machine"	0.62
"Random Forest"	0.60
"AdaBoost"	0.60
"Nearest Neighbors"	0.65
"Decision Tree"	**0.78**

The bold value represents the best model's performance.

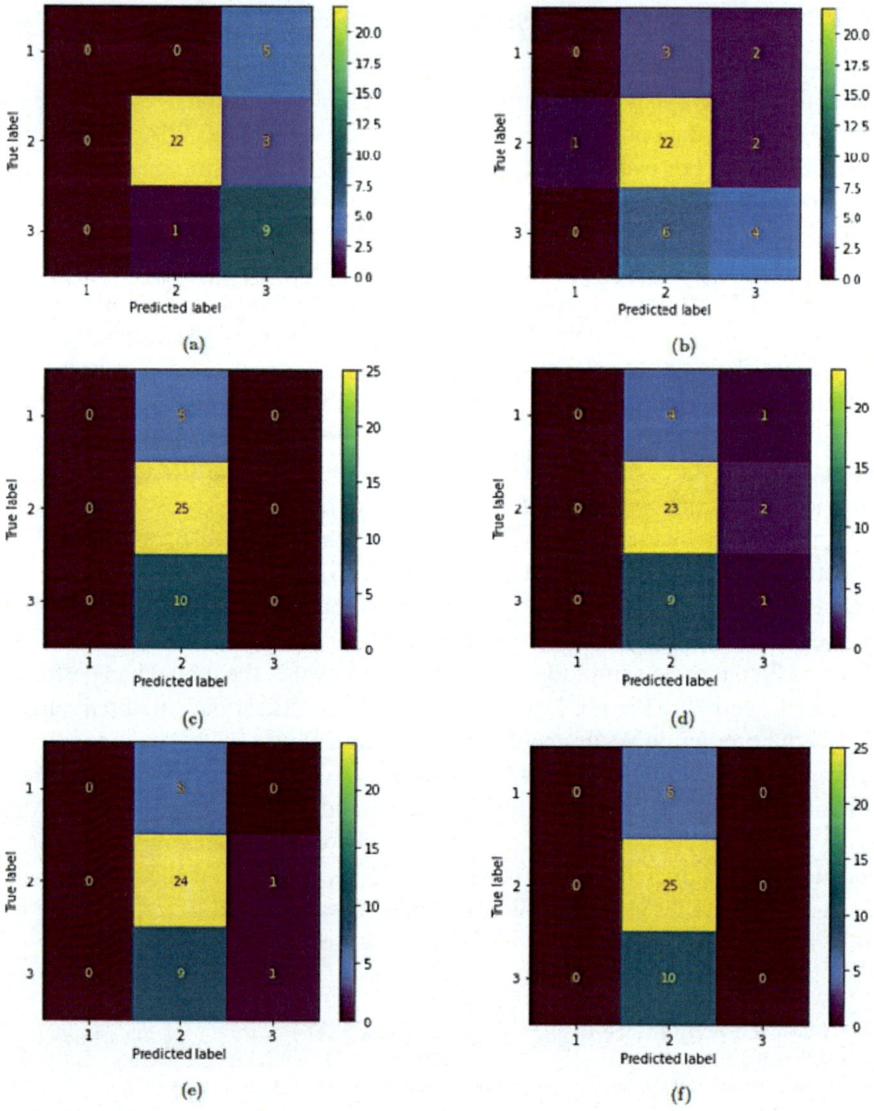

Fig. 12.2 A comparison of six classifiers in scikit-learn on SWP prediction. (**a**) "Decision Trees," (**b**) "Nearest Neighbors," (**c**) "Neural Net," (**d**) "AdaBoost," (**e**) "Random Forest," and (**f**) "Support Vector Machine." A confusion matrix was a summary of prediction results on a classification problem. The number of correct and incorrect predictions was tallied with count values and divided into classes. The confusion matrix provided an insight into not only the errors being made by a classifier but also more importantly the types of errors that were being made. "True label" meant the ground truth of SWP levels. "Predicted label" identified the SWP levels predicted by the trained model

Table 12.4 The "Decision Trees" performance on SWP prediction. In the analysis as "Decision Tree" for the SWP Level 1, the trained model predicted the test data was in the range of Level 3. Therefore, the prediction accuracy (F1-score) of Level 1 was zero. The model successfully classified 22 out of 25 samplings in Level 2. For Level 2, the model had performance with an F1-score of 0.92. For Level 3, the model classified nine out of ten samplings. The overall prediction accuracy of the trained model was 0.78

SWP levels	Precision	Recall	F1-score
Level 1	0	0	0
Level 2	0.96	0.88	0.92
Level 3	0.53	0.90	0.67
Accuracy	NA	NA	0.78
Macro avg	0.50	0.59	0.53
Weighted avg	0.73	0.78	0.74

12.4 Summary

The detection of walnut SWP is important in the walnut industry for management decision support. In this study, a proximate sensor was used to classify different walnut SWP levels. By using the "Decision Tree" classifier, the Walabot could classify the walnut SWP levels with an accuracy of 78%. As a cutting-edge technology for radio sensing, the sensor could be used flexibly in walnut orchards and provided accurate data information. In the future, more data will be collected for training the model. With more data from the walnut trees, more training data will hypothetically improve the accuracy of the model.

References

1. Allen, C.D., Macalady, A.K., Chenchouni, H., Bachelet, D., McDowell, N., Vennetier, M., Kitzberger, T., Rigling, A., Breshears, D.D., Hogg, E.T., et al.: A global overview of drought and heat-induced tree mortality reveals emerging climate change risks for forests. Forest Ecol. Manag. **259**(4), 660–684 (2010)
2. Browne, M., Yardimci, N.T., Scoffoni, C., Jarrahi, M., Sack, L.: Prediction of leaf water potential and relative water content using terahertz radiation spectroscopy. Plant Direct **4**(4), e00,197 (2020)
3. Goldhamer, D., Fereres, E., et al.: Simplified tree water status measurements can aid almond irrigation. Calif. Agricult. **55**(3), 32–37 (2001)
4. Jones, H.: Plants and microclimate: A quantitative approach to environmental plant physiology (2013). https://doi.org/10.1017.CBO9780511845727
5. Lampinen, B., Sibbett, S., Olson, W., Shackel, K.: The relation of midday stem water potential to the growth and physiology of fruit trees under water limited conditions. In: III International Symposium on Irrigation of Horticultural Crops 537, pp. 425–430 (1999)
6. Niu, H., Hollenbeck, D., Zhao, T., Wang, D., Chen, Y.: Evapotranspiration estimation with small UAVs in precision agriculture. Sensors **20**(22), 6427 (2020)
7. Sack, L., John, G.P., Buckley, T.N.: ABA accumulation in dehydrating leaves is associated with decline in cell volume, not turgor pressure. Plant Physiol. **176**(1), 489–495 (2018)

8. Zhao, T., Koumis, A., Niu, H., Wang, D., Chen, Y.: Onion irrigation treatment inference using a low-cost hyperspectral scanner. In: Multispectral, Hyperspectral, and Ultraspectral Remote Sensing Technology, Techniques and Applications VII. International Society for Optics and Photonics (2018)
9. Zhao, T., Yang, Y., Niu, H., Wang, D., Chen, Y.: Comparing U-Net convolutional network with mask R-CNN in the performances of pomegranate tree canopy segmentation. In: Multispectral, Hyperspectral, and Ultraspectral Remote Sensing Technology, Techniques and Applications VII, vol. 10780, p. 107801J. International Society for Optics and Photonics (2018)

Chapter 13
A Low-Cost Soil Moisture Monitoring Method by Using Walabot and Machine Learning Algorithms

Abstract This chapter introduces an economical soil moisture monitoring method utilizing Walabot technology and machine learning algorithms. The chapter initiates with an introduction, emphasizing the importance of cost-effective solutions for soil moisture assessment in digital agriculture. The Materials and Methods section provides an overview of the study site, the utilization of the proximate sensor, and the experimental setup, along with details on data collection and processing. The chapter emphasizes the affordability and accessibility of the Walabot-based approach. The Results and Discussion section presents the outcomes of the experiments, showcasing the performance of linear discriminant analysis and principal component analysis in soil moisture estimation. The chapter concludes with a summary, underlining the potential of this low-cost method for widespread soil moisture monitoring. References are provided for further exploration of the methodologies and technologies discussed in this chapter.

13.1 Introduction

In [3], Engman et al. defined soil moisture as the temporary storage of precipitation within a shallow layer of the Earth. The soil moisture plays an important role in hydrological applications, such as agriculture [21], climate change [4], and meteorology [5]. For example, the data analysis from soil moisture monitoring can be used for crop yield estimation, irrigation treatment inference [11], and warning of drought [3]. Soil moisture monitoring can also be applied for pest management [13] and evapotranspiration estimation [10]. Therefore, it is important to monitor the soil moisture accurately. Typically, there are two types of methods for monitoring the soil moisture, proximate sensing, and remote sensing.

Proximate sensing methods for soil moisture are currently restricted to point-specific measurements [17]. For example, researchers usually put the soil moisture probes in the test field for monitoring. However, these discrete measurements cannot represent the spatial and temporal soil moisture distribution for the whole field.

With the development of remote sensing technology, the satellite has been widely used for soil moisture remote sensing [2]. Many researchers have proved that

H. Niu, Y. Chen, *Smart Big Data in Digital Agriculture Applications*, Agriculture Automation and Control, https://doi.org/10.1007/978-3-031-52645-9_13

optical and thermal remote sensing can be used for soil moisture measurements. For example, in [16], Wang et al. proposed the normalized multiband drought index (NMDI) for remotely sensing the soil based on the soil spectral characteristic. Since variations of soil moisture have a significant influence on soil surface temperature [6], thermal infrared remote sensing is also used for measuring the soil temperature to correlate it with soil moisture. Active and passive microwave remote sensing techniques are also commonly used for soil moisture measurements [15]. For passive microwave sensors, they can measure the intensity of microwave emission from the soil, which is proportional to the brightness temperature, a product of the surface temperature and emissivity [17, 19]. However, there are disadvantages to these methods. Limited surface penetration can be a problem for both optical and thermal remote sensing. Cloud contamination can be another issue [17]. The data acquired from the microwave have a low spatial resolution.

Therefore, in this study, the authors proposed a new low-cost (less than $ 1000) soil moisture monitoring method by using a Walabot sensor and machine learning algorithms. Walabot is a pocket-sized device and cutting-edge technology for radio frequency tridimensional sensing, which has already been used in many research topics, such as nematodes detection [12] and battery voltage detection [18]. It can work flexibly in the field and provide data information more accurately than remote sensing methods with machine learning algorithms. First, the sensor was used to collect radio frequency reflectance of sampling soil, as shown in Fig. 13.1, which can detect the physical structure of the soil moisture. Second, the collected data were preprocessed by data enhancement or a wavelet transform. Third, processed data were used by PCA [20] and LDA [1] for analysis. Results showed that the Walabot successfully classified the different levels of soil moisture with a state-of-the-art performance. Moreover, with the development of wireless technology

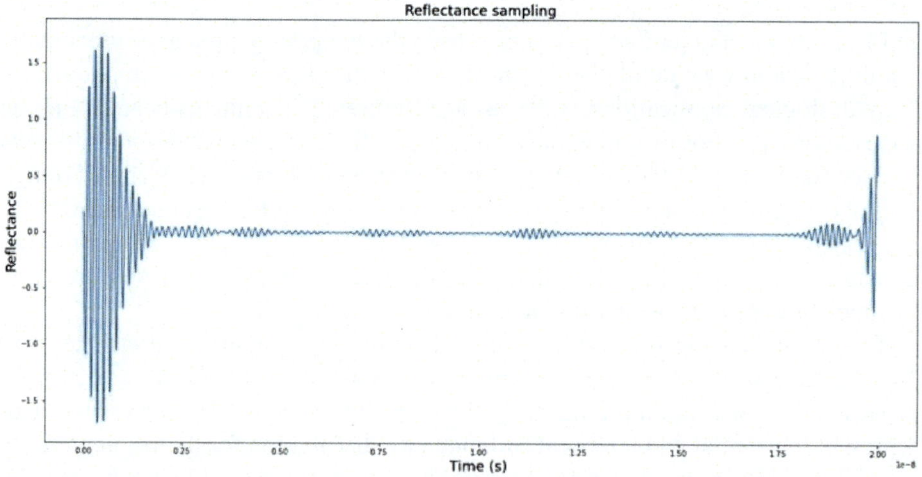

Fig. 13.1 Radio Frequency reflectance. The Walabot Developer is a programmable 3D sensor that uses radio frequency to see through the soil and creates a reflectance image within 1 second

and microelectromechanical systems and computer vision, we might even use the Walabot to recognize real-time soil moisture monitoring in the future research.

13.2 Materials and Methods

13.2.1 The Study Site

This research was conducted at the Mechatronics, Embedded Systems, and Automation (MESA) Lab in Atwater, CA, USA ($37°22'30.6''N$, $120°34'40.9''W$).

13.2.2 The Proximate Sensor

The sensor being used was Walabot Developer (Vayyar Imaging Ltd), as shown in Fig. 13.2. The Walabot Developer is a programmable 3D sensor that uses radio frequency to see through the soil and creates a reflectance image within 1 second. The frequency range is 3.3–10 GHz (US/FCC model) and 6.3–8 GHz (EU/CE model). The average transmit power of both models is below 41 dBm/MHz and

Fig. 13.2 Walabot data collection. The experiment was conducted in the MESA Lab. The soil was sampled in an almond field near the lab and was divided into three cups

does not have any health problem. In principle, the Walabot uses an antenna array to illuminate the area in front of it and captures the returning signals. The signals are produced and recorded by VYYR2401 A3 System-on-Chip integrated circuits.

Based on the technical specs, the Walabot can sense the environment by transmitting, receiving, and recording signals from multiple antennas. Multiple transmit–receive antenna pairs' recordings are analyzed to build a 3D image of the environment. Then, researchers can detect changes in the environment by analyzing the sequences of images. The sensor is also capable of short-range imaging into dielectric environments, such as drywall and concrete. Therefore, it can be used in many study areas as follows:

1. In room/wall imaging
2. Object detection, location, and tracking
3. Speed measurement and motion sensing
4. Dielectric properties of materials sensing

13.2.3 Experiment Setup

In this study, the authors used Walabot to detect different levels of soil moisture. The experiment was conducted in the MESA Lab. The soil was sampled in an almond field near the lab and was divided into three cups, as shown in Fig. 13.2. All the soil samplings were from the same spot in the almond field to make sure they are homogeneous. The soil was dried out to make sure all the three cups of soil were at the same lowest moisture level. The weights of three cups of dry soil were 632 grams, 630 grams, and 634 grams. 6 g or 8 g of water was added in every cup each time (ten times in total) to increase the soil moisture until the soil was saturated, as shown in Table 13.1.

13.2.4 Data Collection and Processing

The Walabot was used to measure the soil moisture every time after the water was added. Each measurement by the Walabot was repeated ten times to reduce the likelihood of errors or anomalous results so that it could increase the confidence interval.

For image processing, the authors used two different machine learning methods, the Principal Component Analysis (PCA) and Linear Discriminant Analysis (LDA). Both of them could reduce the dimensionality of the datasets and increase the classification accuracy.

Table 13.1 Soil samplings. All the soil samplings were from the same spot in the almond field to make sure they are homogeneous. The soil was dried out to make sure all the three cups of soil were at the same lowest moisture level. The weights of three cups of dry soil were 632 grams, 630 grams, and 634 grams. 6 g or 8 g of water was added in every cup each time (ten times in total) to increase the soil moisture until the soil moisture was saturated

Soil condition	Soil sample 1	Soil sample 2	Soil sample 3
Dry	632g	630g	634g
1	640g	638g	642g
2	648g	646g	648g
3	654g	652g	656g
4	660g	660g	664g
5	668g	668g	670g
6	676g	674g	678g
7	682g	682g	686g
8	690g	690g	692g
9	696g	696g	700g
10 (Saturated)	704g	704g	708g

13.2.4.1 Principal Component Analysis

Principal Component Analysis (PCA) is a fast and flexible unsupervised method for data dimensionality reduction [8]. It can achieve linear projection to a lower dimensional subspace by using singular value decomposition. It is a linear transformation that rotates the axes of the data along the direction that maximizes its variance, allowing data to be projected into a lower dimensional subspace [8]. These new axes, or "loadings," are found by calculating the eigenvectors W of the covariance matrix of the data, where X was an $M \times N$ matrix representing M samples of size N

$$X^{\mathrm{T}} X = W\lambda. \tag{13.1}$$

The eigenvalue λ represents the importance of the loading in transforming the data. As the loadings are sorted in descending order, W can be truncated to r columns, which can then be used to project data along r dimensions, preserving the dimensions that contribute most to the variance of the distribution. W is often obtained with Singular Value Decomposition (SVD) instead of performing the eigen decomposition of $X^{\mathrm{T}} X$, as it is more computationally efficient.

PCA can also maximize the variance of the projected data. Therefore, PCA is commonly used in exploratory data analysis and making predictive models.

13.2.4.2 Linear Discriminant Analysis

Linear Discriminant Analysis (LDA) is a classifier with a linear decision boundary. It is generated by using Bayes' rule to fit class conditional densities to the data. It assumes that all classes share the same covariance matrix. After that, the LDA model can be used to reduce the dimensionality of the input data by projecting it to the most discriminative directions. The model will maximize the distance between means of classes relative to some center point for all classes, while minimizing the variance, or scatter, within each category. In the following equation, C is the number of classes, N_i is the size of class i, μ is the mean of all data points, μ_i is the mean of class i, and x_j is the jth data point in class i

$$\frac{\sum_{i=1}^{C} N_i(\mu_i - \mu)(\mu_i - \mu)^{\mathrm{T}}}{\sum_{i=1}^{C}\sum_{j=1}^{N_i}(x_j - \mu_i)(x_j - \mu_i)^{\mathrm{T}}}. \tag{13.2}$$

The optimized solution contains eigenvectors, which are in descending order per their eigenvalues, which can be used to reduce the dataset similar to PCA. Optimizing for both within- and between-class scatter is essential because only maximizing distance between means can lead to scenarios where the variance is high along the axis with great mean distances, increasing the chance that there are points from different classes overlapping. Minimizing the variance ensures that data from each class are grouped tightly along the new axis, increasing separability. Then, the output dimensionality is usually less than the number of classes, so that LDA is a very strong dimensionality reduction [7].

13.3 Results and Discussion

Each radio frequency reflectance image was converted into a 2048-dimensional vector for data processing. The data were distributed as 67% for training and 33% for testing. Since the dataset was small for training 11 classifiers, the authors distributed the eleven soil conditions into five different levels from dry to saturation. As shown in Figs. 13.3 and 13.4, "Dry" meant the dry soil. "WetTotal" stands for the saturated soil. Soil conditions 1, 2, and 3 were included in "Wet1." "Wet2" contained the soil conditions 4, 5, and 6. "Wet3" included the soil conditions 7, 8, and 9.

Several classifiers in scikit-learn were used for comparison, such as "Nearest Neighbors," "Linear SVM," "RBF SVM," "Gaussian Process," "Decision Tree," "Random Forest," "Neural Net," "AdaBoost," "Naive Bayes," and "QDA." In this soil moisture monitoring problem, the accuracy of these classifiers is shown in Table 13.2. The best classifiers were "Nearest Neighbors," "Gaussian Process," "Decision Tree," "Random Forest," "Neural Net," and "Naive Bayes" with an accuracy of 95%. The "QDA" was with 90% accuracy. The "Linear SVM" and "AdaBoost" were worst with 55% accuracy.

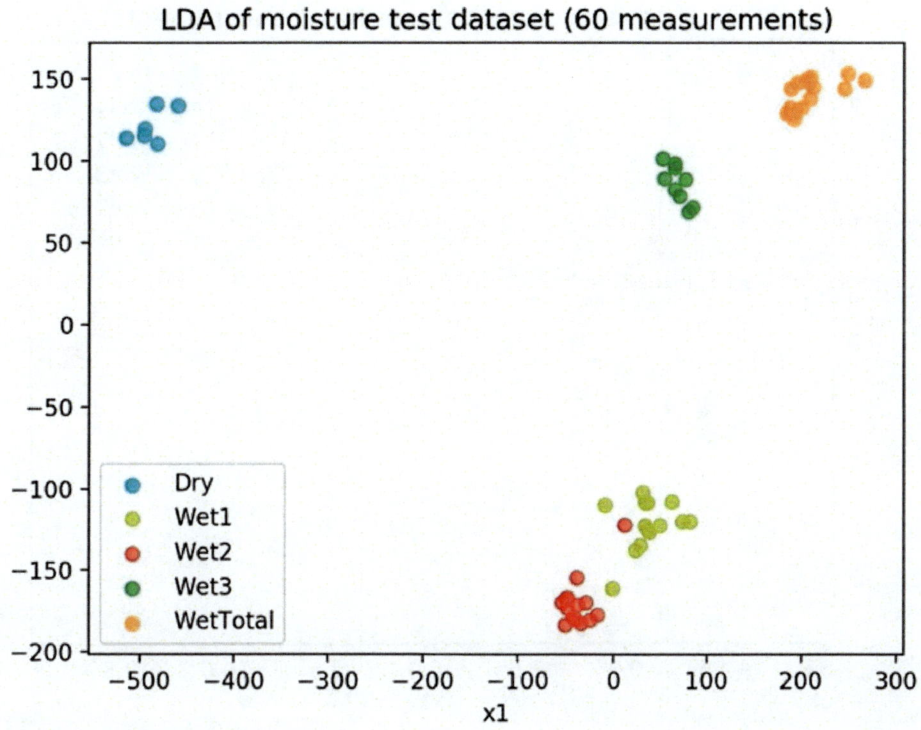

Fig. 13.3 LDA results for soil moisture measurement. There were five different soil moisture levels with different colors, "Dry," "Wet1," "Wet2," "Wet3," and "WetTotal," which meant the soil sampling was saturated. LDA classifiers first reduced the original dimension to two components. Different colors meant different soil moisture levels and the axes of the figure were dimensionless. The LDA could classify the five different levels of soil moisture in different areas of the coordinate, so that the LDA could classify the soil moisture with an state-of-the-art performance

Scikit-learn's accuracy classification score function evaluated the performance of the classifiers. This function computed the subset accuracy, in which the labels predicted for sampling must exactly match the corresponding true labels. Estimators used this score method as the evaluation criterion for the classification problems. All scorer objects followed the convention that higher return values were better than lower return values.

13.3.1 Linear Discriminant Analysis Performance

Several LDA methods were used for soil moisture classification, as shown in Table 13.3. decision_function(X) was for predicting confidence scores of soil samples. fit(X, y[, store_covariance, tol]) was for fitting the LDA model according to the given soil image training data and parameters. fit_transform(X[, y]) was for fitting data and transform it. get_params([deep]) is used for setting parameters

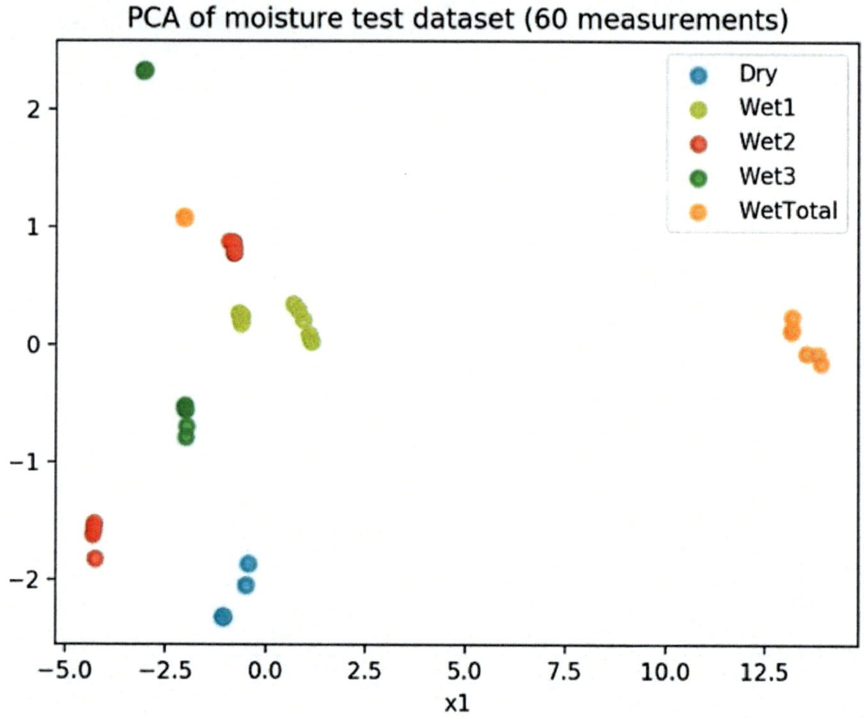

Fig. 13.4 PCA results for soil moisture measurement. "WetTotal" points were on the left and right sides of the image. The "Wet3" and "Wet2" data points did not drop in the same area. The reason might be that the PCA could not detect the feature difference from the data. Similar to LDA, the PCA classifiers first reduced the original dimension to two components. Then, each classifier was tested against reduced dimensionality data with the component as two

for the estimator. Then, predict(X) could predict the class labels for soil samples. predict_log_proba(X) and predict_proba(X) could estimate the probability. Finally, score(X, y[, sample_weight]) could return the mean accuracy on the given soil images test data and labels. set_params(**params) was for setting estimator parameters. transform(X) was for projecting data to maximize soil class separation.

The performance of the LDA for soil moisture monitoring is shown in Fig. 13.3. There were five different soil moisture levels with different colors, "Dry," "Wet1," "Wet2," "Wet3," and "WetTotal," which meant the soil sampling was saturated. LDA classifiers first reduced the original dimension to two components. As seen from Fig. 13.3, different colors meant different soil moisture levels and the axes of the figure were dimensionless. The LDA could classify the five different levels of soil moisture in different areas of the coordinate, so that the LDA could classify the soil moisture with a state-of-the-art performance.

Table 13.2 Classifiers accuracy. The best classifiers were "Nearest Neighbors," "Gaussian Process," "Decision Tree," "Random Forest," "Neural Net," and "Naive Bayes" with an accuracy of 95%. The "QDA" was with 90% accuracy. The "Linear SVM" and "AdaBoost" were worst with 55% accuracy

Classifiers	Accuracy
"Nearest Neighbors"	0.95
"Linear SVM"	0.40
"RBF SVM"	0.95
"Gaussian Process"	0.95
"Decision Tree"	0.95
"Random Forest"	0.95
"Neural Net"	0.95
"AdaBoost"	0.55
"Naive Bayes"	0.95
"QDA"	0.90

Table 13.3 The LDA methods. Several LDA methods were used for soil moisture classification

Methods
decision_function(X)
fit(X, y[, store_covariance, tol])
fit_transform(X[, y])
get_params([deep])
predict(X)
predict_log_proba(X)
predict_proba(X)
score(X, y[, sample_weight])
set_params(**params)
transform(X)

13.3.2 Principal Component Analysis Performance

In PCA methods, fit(self, X[, y]) was to fit the model with the input soil images data X. fit_transform(self, X[, y]) was for fitting the model with X and applying the dimensionality reduction on X. get_covariance(self) was for computing the data covariance with the generative model. get_params(self[,deep]) was to get parameters for the estimator. get_precision(self) was for computing the data precision matrix with the generative model. Then, inverse_transform(self, X) could transform the data back to its original space. Finally, score(self, X[, y]) could return the average log-likelihood of all samples. score_samples(self, X) could return the log-likelihood of each sample. set_params(self, params) could help set the parameters of the estimator. transform(self, X) was being used for applying dimensionality reduction to soil images input (Table 13.4).

In Fig. 13.4, PCA could also classify the soil moisture successfully but not entirely. As shown in Fig. 13.4, "WetTotal" points were on the left and right sides of the image. The "Wet3" and "Wet2" data points did not drop in the same area. The reason might be that the PCA could not detect the feature difference from the data. Similar to LDA, the PCA classifiers first reduced the original dimension to two components. Then, each classifier was tested against reduced dimensionality

Table 13.4 The PCA methods

Methods
fit(self, X[, y])
fit_transform(self, X[, y])
get_covariance(self)
get_params(self[,deep])
get_precision(self)
inverse_transform(self, X)
score(self, X[, y])
score_samples(self, X)
set_params(self, **params)
transform(self, X)

data with the component as two. Results showed that LDA performs much better than the PCA method.

13.4 Summary

Soil moisture monitoring is essential in digital agriculture, which has a significant effect on crop evapotranspiration, the exchange of water, and energy fluxes. Soil moisture can be measured by many remote sensing or proximate sensing techniques, such as thermal, optical, and microwave measurements. However, there are limiting factors for the applications of these methods, such as low spatial resolution, limited surface penetration, and vegetation. In this study, the authors used a portable sensor to classify different soil moisture successfully. By using the PCA and LDA machine learning methods, the Walabot could recognize small changes in different levels of soil moisture and can detect soil moisture difference with an accuracy of 100% by LDA algorithms, which showed a state-of-the-art performance in estimating soil moisture. As a pocket-sized device cutting-edge technology for radio frequency tridimensional sensing, the authors believe the sensor can work flexibly in the field and provide data information more promptly and accurately than traditional remote sensing or proximate sensing method.

So far, the Walabot can only detect the difference in soil moisture. In the future, the authors will compare it with different soil moisture sensors to see if we can find the regression model and quantify the soil moisture measurements by using Walabot. With the development of wireless technology and microelectromechanical systems and computer vision, it might be able to be mounted on unmanned ground vehicles (UGVs) for proximate sensing [9, 14]. Thus, we can use the sensor to recognize real-time soil moisture for continuous yet possibly mobile monitoring.

References

1. Balakrishnama, S., Ganapathiraju, A.: Linear discriminant analysis: A brief tutorial. Inst. Signal Inf. Process. **18**, 1–8 (1998)
2. Engman, E.T.: Progress in microwave remote sensing of soil moisture. Canad. J. Remote Sens. **16**(3), 6–14 (1990)
3. Engman, E.T.: Applications of microwave remote sensing of soil moisture for water resources and agriculture. Remote Sens. Environ. **35**(2-3), 213–226 (1991)
4. Entekhabi, D., Nakamura, H., Njoku, E.G.: Retrieval of soil moisture profile by combined remote sensing and modeling. In: Passive Microwave Remote Sensing of Land–Atmosphere Interactions, pp. 485–498. CRC Press (2023)
5. Fast, J.D., McCorcle, M.D.: The effect of heterogeneous soil moisture on a summer baroclinic circulation in the central United States. Month. Weather Rev. **119**(9), 2140–2167 (1991)
6. Friedl, M., Davis, F.: Sources of variation in radiometric surface temperature over a tallgrass prairie. Remote Sens. Environ. **48**(1), 1–17 (1994)
7. Friedman, J., Hastie, T., Tibshirani, R.: The Elements of Statistical Learning, vol. 1. Springer Series in Statistics, New York (2001)
8. Jolliffe, I.: Principal Component Analysis. Springer, New York (2011)
9. Niu, H., Zhao, T., Chen, Y.: Intelligent bugs mapping and wiping (iBMW): An affordable robot-driven robot for farmers. In: IEEE International Conference on Mechatronics and Automation (ICMA), pp. 397–402. IEEE (2019)
10. Niu, H., Zhao, T., Wang, D., Chen, Y.: Estimating evapotranspiration with UAVs in agriculture: A review. In: Proceedings of the ASABE Annual International Meeting. American Society of Agricultural and Biological Engineers (2019)
11. Niu, H., Zhao, T., Wang, D., Chen, Y.: A UAV resolution and waveband aware path planning for onion irrigation treatments inference. In: 2019 International Conference on Unmanned Aircraft Systems (ICUAS), pp. 808–812. IEEE (2019)
12. Niu, H., Zhao, T., Westphal, A., Chen, Y.: A low-cost proximate sensing method for early detection of nematodes in walnut using Walabot and scikit-learn classification algorithms. In: Autonomous Air and Ground Sensing Systems for Agricultural Optimization and Phenotyping V. International Society for Optics and Photonics (2020)
13. Powell, L., Berg, A.A., Johnson, D., Warland, J.: Relationships of pest grasshopper populations in Alberta, Canada to soil moisture and climate variables. Agricult. Forest Meteorol. **144**(1-2), 73–84 (2007)
14. Tian, J., Niu, H., Wang, P., Chen, Y.: Smart and autonomous farm field scouting service robot as an edge device under $1000: Challenges and opportunities. In: ASME 2019 International Design Engineering Technical Conferences and Computers and Information in Engineering Conference. American Society of Mechanical Engineers Digital Collection (2019)
15. Walker, J.P.: Estimating soil moisture profile dynamics from near-surface soil moisture measurements and standard meteorological data. Ph.D. thesis, University of Newcastle (1999)
16. Wang, L., Qu, J.J.: NMDI: A normalized multi-band drought index for monitoring soil and vegetation moisture with satellite remote sensing. Geophys. Res. Lett. **34**(20) (2007)
17. Wang, L., Qu, J.J.: Satellite remote sensing applications for surface soil moisture monitoring: A review. Front. Earth Sci. China **3**(2), 237–247 (2009)
18. Wang, Y., Niu, H., Zhao, T., Liao, X., Dong, L., Chen, Y.: Contactless Li-Ion battery voltage detection by using Walabot and machine learning. In: Proceedings of the ASME 2019 International Design Engineering Technical Conferences and Computers and Information in Engineering Conference. American Society of Mechanical Engineers Digital Collection (2019)
19. Wigneron, J.P., Calvet, J.C., Pellarin, T., Van de Griend, A., Berger, M., Ferrazzoli, P.: Retrieving near-surface soil moisture from microwave radiometric observations: current status and future plans. Remote Sens. Environ. **85**(4), 489–506 (2003)

20. Wold, S., Esbensen, K., Geladi, P.: Principal component analysis. Chemomet. Intell. Lab. Syst. **2**(1-3), 37–52 (1987)
21. Zhao, T., Koumis, A., Niu, H., Wang, D., Chen, Y.: Onion irrigation treatment inference using a low-cost hyperspectral scanner. In: Multispectral, Hyperspectral, and Ultraspectral Remote Sensing Technology, Techniques and Applications VII. International Society for Optics and Photonics (2018)

Chapter 14
Conclusion and Future Research

14.1 Concluding Remarks

This book presented the concept and importance of smart big data toward digital agricultural applications. Smart big data in agricultural applications is an interdisciplinary research topic related to the extraction of meaningful information from plant physiology data, drawing techniques from a variety of fields, such as UAV image processing, deep learning, pattern recognition, high-performance computing, and statistics. Big data can then be filtered and becomes smart big data before being analyzed for insights, which leads to more efficient decision making. Smart big data can be defined as big data that has been (1) collected, (2) preprocessed (such as cleaned, filtered, and prepared for data analysis), and (3) actionable information extraction in a smart way. Likewise, the authors proposed the smart big data acquisition platforms, such as small UAVs and Edge-AI sensors. Smart big data cleans and transforms the data into useful information that is valuable and relevant to crops and trees growing status.

The authors proposed the concept of plant physiology-informed machine learning and how to use advanced analytics and fractional order thinking to make contributions. Numerous inexperienced researchers in the domain of digital agriculture frequently encounter difficulties when attempting to leverage AI for agricultural purposes, particularly if they have limited computer science expertise. Mastering the complexities of machine learning and computer vision can be a time-consuming and daunting endeavor. To address this prevalent challenge, we have allocated Chap. 6 to a user-friendly tutorial tailored for beginners. This tutorial comprehensively covers the foundational components of a machine learning or computer vision workflow. Its purpose is to furnish a robust foundation and facilitate the initiation of individuals with minimal prior familiarity with AI-driven agricultural solutions. In Chap. 7, a non-invasive proximate sensing method for early detection of nematodes was proposed. Microwave reflectance from walnut leaves was analyzed using ML algorithms to classify the nematode infection levels in the walnut roots. In Chap. 8,

H. Niu, Y. Chen, *Smart Big Data in Digital Agriculture Applications*, Agriculture Automation and Control, https://doi.org/10.1007/978-3-031-52645-9_14

reliable tree-level ET estimation methods were proposed using the UAV high-resolution imagery, ML algorithms, and platforms, such as Python, MATLAB, PyTorch, and TensorFlow. In Chap. 9, individual tree-level water status inference was performed using the high-resolution UAV thermal imagery and complexity-informed machine learning. In Chap. 10, the authors proposed a scale-aware pomegranate yield prediction method using UAV imagery and machine learning. Then, the authors discussed an intelligent bugs mapping and wiping robot for farmers in Chap. 11, which has great potential for pest management in the future. Then, the authors proposed a non-invasive stem water potential monitoring method using proximate sensor and ML algorithms for a walnut orchard in Chap. 12 and a low-cost soil moisture monitoring method in Chap. 13.

Smart big data is the use of various methods, such as ML algorithms and artificial intelligence, to analyze and transform the agricultural data into information from which valuable insight can be drawn. It incorporates advanced analytics to enhance plant physiology-informed machine learning. The application of smart big data can help researchers and stakeholders develop a better understanding of the data with the goal of developing better digital agriculture. So far, these are still initial steps for the development of the smart big data applications in digital agriculture. More development of the smart big data framework integrated with advanced analytics, ML algorithms, and AI is required to demonstrate the capabilities of this innovative methodology.

14.2 Future Research Toward Smart Big Data in Digital Agricultural Applications

Based on the research presented, a comprehensive work has been done toward the smart big data in digital agricultural applications. There is still plenty of work left to be done. To our current thinking, the future research plans will be:

- A new book project plan: "AI in Digital Agriculture: A Cookbook Guide"
- Further investigate the necessity of smart big data and the correlation between smart big data, machine learning, and plant physiology.
- Design and test a UGV platform for agricultural applications, such as pest population distribution, quantification, and management. Develop an intelligent bugs mapping and wiping system using UGVs for pest management.
- Evaluate the remote sensing methods of early detection of nematode using small UAVs and lightweight sensors, such as SWIR, TIR, and multispectral cameras. Combine the proximate sensing methods and remote sensing methods to generate a comprehensive framework for early detection of nematodes.
- Develop a computer vision-based nematode counting and segmentation method using ML algorithms for nematode image processing.
- Investigate TSEB and DTD methods for reliable tree-level ET estimation using small UAVs and remote sensing sensors.

- Demonstrate the importance of spatial and temporal analysis for tree-level yield estimation using small UAVs and remote sensing sensors.
- Develop a robust and adaptive individual tree-level irrigation treatment inference system using high-resolution UAV thermal imagery and complexity-informed machine learning.
- Define and implement the concept of complexity-informed machine learning and principle of tail matching with Edge-AI sensors for more agricultural applications, such as soil moisture monitoring and stem water potential measurement.
- Further improve the concept of IoLT and demonstrate its key role with ML and Edge-AI sensors.

In summary, big data has to be smart so that smart actions become possible. Artificial intelligence in digital agriculture has to respect and respond to the complexity in the plant physiological processes.

Index

H. Niu, Y. Chen, *Smart Big Data in Digital Agriculture Applications*, Agriculture Automation and Control, https://doi.org/10.1007/978-3-031-52645-9